An Engineer's Guide to Technical Communication

Sheryl A. Sorby
Michigan Technological University

William M. Bulleit
Michigan Technological University

PEARSON
Prentice Hall

Upper Saddle River, New Jersey
Columbus, Ohio

Library of Congress Cataloging-in-Publication Data

Sorby, Sheryl Ann
  An engineer's guide to technical communication / Sheryl A. Sorby, William M. Bulleit.
     p. cm.
  Includes bibliographical references and index.
  ISBN 0–13–048242–0
  1. Communication in engineering. 2. Technical writing. I. Bulleit, William M. II. Title.

  TA158.5.S65 2006
  808'.0666—dc22   2004029647

Editor: Gary Bauer
Editorial Assistant: Jacqueline Knapke
Production Editor: Kevin Happell
Design Coordinator: Diane Ernsberger
Cover Designer: Jason Moore
Cover Art: Digital Vision
Production Manager: Pat Tonneman
Marketing Manager: Ben Leonard

This book was set in Stone Serif by *The GTS Companies*/York, PA Campus. It was printed and bound by Courier Stoughton, Inc. The cover was printed by The Lehigh Press, Inc.

Pearson Education Ltd.
Pearson Education Singapore Pte. Ltd.
Pearson Education Canada, Ltd.
Pearson Education—Japan

Pearson Education Australia Pty. Limited
Pearson Education North Asia Ltd.
Pearson Educación de Mexico, S.A. de C.V.
Pearson Education Malaysia Pte. Ltd.

10 9 8 7 6 5 4 3 2 1
ISBN 0-13-048242-0

To all of the tough women in my life who inspired me to go for it—
Ethel, Flora, Irene, Mariam, Carrie, and Claire.

To Gladys Youngs, who taught thousands of Hastings High students
to write effectively (whether they wanted to or not)
and who started me on this path.

—S.A.S

To my wife, Laura, who keeps my head from getting too big, and to my
children, Brittany, Carly, and Clayton, who keep me on my toes.

To all of those people who said that engineers can't write—thanks for
making me want to prove you wrong.

—W.M.B.

Contents

Preface

Everyone knows that engineers* use mathematical and scientific principles to solve problems and to design solutions. To many people, the stereotypical engineer is an introvert who rarely interacts with others and who enjoys spending time deriving equations and integrating functions. Although engineers usually do have excellent math and science skills, and some of us even enjoy working with complex mathematical functions, what is less well known is that engineers must have excellent communication skills to be effective professionals. Most practicing engineers will readily admit they spend more time on communication tasks than on "real" engineering. As a person moves up through the ranks, assuming positions of leadership in a company, well-developed communication skills become even more critical to success. This text was written with that fact in mind. Similar to most other types of skills, effective communication skills are best developed through practice. The more you work at it, the better you become. Although you may not want to do so, you should keep working on your communication skills throughout your college career and beyond.

In engineering design, one well-known maxim is "Form follows function." What this saying means is that when an engineer sets out to design a system, the function the system will serve dictates what the final result will look like. We have taken this approach to heart in writing this text. Every type of technical communication has a specific function within the engineering profession, and the form the communication takes often depends on its function. For each type of communication described in this text, we first tell you its *function* and then its *form*. We follow these by a section on *mechanics*. The purpose of the mechanics section is to provide details about exactly how a certain type of communication is to be created. The details help you achieve the overall form that is required. Following the sections on function, form, and mechanics for each type of communication are examples you can use as guides during preparation of your own technical documents.

---

*In this text, *engineers* is used to signify both engineers and engineering technologists. Further, our definition of engineers includes all types—operating engineers, test engineers, design engineers, and so forth.

Many texts are available that focus on technical communication or on technical writing; however, we believe this text is decidedly different from most of these other texts. This text was written *by* engineers, *for* engineers. As engineers, we (the authors) are in the business of communicating technical information daily. We do not merely *study* technical communication, we *live* it.

Although we are sure you will find many differences between this text and others that focus on technical communication, we would like to point out some of the specific features that we believe set this text apart from other texts you may encounter:

- A section on style and grammar is included early in the text (in Chapter 2), and the Appendix provides more extensive coverage of this topic. *Poor grammar and style detract from the message you are trying to convey.* Just as you cannot solve calculus problems without understanding the basic rules of algebra and trigonometry, you will not be able to communicate technical information effectively if you do not understand the basic rules of English grammar. Although no one expects your grammar to be flawless, you should know enough about grammar so that others will respect your work and think you are actually a college graduate.

- Unlike other texts, this text includes a chapter on communication of design solutions through calculation sheets. Design calculation documentation is a vital part of technical communication. Calculation documentation is a significant part of experimental, design, and other types of technical reports. As a practicing engineer, you must properly document your calculations for many reasons. If your work will be reviewed by another engineer or a colleague, she must be able to understand what you did and the assumptions you made. If one of your projects becomes the focus of litigation, the documentation of your calculations could mean the difference between a judgment of negligence and a judgment of no fault.

- To engineers, communication by graphical means is probably nearly as important as communication through the written word. For this reason, an entire chapter in this text is devoted to the accepted methods for communicating through visual means. Most other technical communication texts we reviewed cover graphical communication briefly, but, in our opinion, do not sufficiently emphasize this critically important aspect. When most engineers (and most of your professors) receive a technical document to review, they first read the abstract or summary to learn about the overall content of the document. As they read through the details, they most likely skim through long sections of text and concentrate on the graphs and figures within the document. If they do not understand a figure or if they want to know more about it, they probably go back and read that section of the text more closely. *This fact does not mean your written word is unimportant,* but it does show the relative importance of your ability to communicate through visual means.

In writing this text, we were forced to organize it into chapters, sections, and subsections. We had many possibilities to choose from. Should we organize around different types of documents? Should we organize around types of communication? Should we organize around types of tools? For example, when we discuss e-mail, should this information go in a section labeled "Communicating Person to Person," or should it go in a chapter titled "Communication in the Electronic Age"? As with most engineering projects, no single solution was correct. The problem in organizing this text was compounded in that many of the topics in the text are interwoven and difficult to categorize in nice, neat packages. In the end, we selected an organization that we were comfortable with and that made sense to us. When topics overlap, references are made to other chapters in the text.

The first chapter of the text is a brief overview of the importance of communication for engineers. We realize this point is a thread that is repeated several times throughout the text, but we think that if we say it often enough, you might actually believe it. As Kipling said, "I've said it thrice; it must be so."

The second chapter is titled "Setting the Stage." In this chapter, we discuss things such as the writing process, tool selection, outlining, grammar, and style. All of these topics are important considerations you must think about *before* you try to communicate technical information. We have all probably experienced college lectures that were unorganized, over our heads, unmotivating, and just plain boring. Thinking about communication tasks before you embark on them will help you to avoid the same mistakes. Also included in Chapter 2 is a section on audience analysis. Consideration of the audience is critically important for effective communication.

Ethical issues in technical communication are covered in Chapter 3. Also included in Chapter 3 are guidelines for searching for and citing references within your technical documents.

Chapter 4 focuses on the various types of personal communication you will prepare in your engineering career. The types of personal communication covered in this chapter include letters, memos, and e-mail.

Chapter 5 covers document design, including sections on tool selection, fonts, and white space. Common document features such as title pages, tables of contents, and appendixes are also covered in this chapter.

Chapter 6 is the heart of the text. When you read this chapter, you will learn about the various written documents you may be asked to prepare during the next 40 years or so. We realize you probably don't believe us when we tell you that you will likely be writing just as much as you will be calculating in your future career (we never believed our engineering professors when they told us this either); but, trust us, this will be the case.

Chapter 7 includes guidelines for documenting your calculations and design work; you will also be preparing this type of documentation during the next 40 years or so. As engineering professors, we can state with certainty that students who organize their calculations neatly and logically on engineer's paper receive higher grades on assignments and tests than those who don't. If we can't determine what you did, even if you did it correctly, chances are that you will lose a significant number of points. If we can see clearly what you did in solving a problem but notice that you made a minor mistake in unit conversions, you will earn nearly all the points for that problem. You will also leave us with a favorable impression of your intelligence, your organizational skills, and your basic understanding of the topic.

Chapter 8 prepares you to communicate orally, specifically to give a presentation in front of a group of classmates or a more formal audience. Speaking in front of an audience may be a frightening experience for you, but you will find that you will be called on to give presentations with surprising frequency. Presentations are common probably because your boss or instructors really don't want to read your report; they just want you to tell them the highlights and provide an opportunity for them to ask questions as needed. They can read it later if necessary.

Chapter 9 covers various types of charts, graphs, and figures used for visual communication. Unlike other forms of communication discussed in the text, this type of communication does not generally stand alone as a document. Rather, visual communication is incorporated into all other forms of communication you may be asked to create. Although graphical images do not typically stand alone as separate documents, they should be able to stand alone within a document. Generally, a person should be able to look at the image alone and understand the meaning you are trying to convey.

Most engineering students want to become gainfully employed in industry at the end of their studies, unless they want to become engineering professors like us and become permanent fixtures on a university campus. Chapter 10 describes two types of special documents you will need to prepare so that you can find a job and start paying back your student loans: résumés and associated cover letters. Although letters are covered in Chapter 4, you should follow special considerations when writing a cover letter that will be attached to your résumé. These considerations are covered in this chapter.

Throughout the text, we offer insight into points you must consider when you are communicating in the electronic age. Remember this: The computer doesn't change the importance of technical communication, and it won't do your job for you. It will not always make your life easier, and in some cases it will make it much more difficult. The value of a computer is always in editing and not in creating. Trust us, a blank computer screen is just as intimidating to us as engineering professors as a blank sheet of paper in a typewriter was to us more than 20 years ago as engineering students.

Acknowledgments

We included several examples throughout this text. When feasible, we tried to include examples from different engineering disciplines. Several students and colleagues at Michigan Tech supplied us with documents and figures that we used to create some of the examples in this text. Specifically, we would like to thank the following individuals:

Undergraduate students Melzar Coulter, Clayton Cooke, Brian Szwejkowski, and Kyle Rubin for figures used in the assembly instructions example (Figure 6.9)

Professor Warren Perger for the use of a document that was the basis for the proposal example (Figure 6.4)

Undergraduate students Jonathan Gohl, Nick Bevins, Salil Kamat, and Todd Wehner, whose senior design project report was the basis for the design project report example (Figure 6.5)

Graduate student Scott Meirs, who graciously allowed us to use a figure from his thesis to illustrate schematic drawings (Figure 9.14A)

Our friend and colleague Associate Professor Tess Ahlborn, who shared a particularly lovely example of a photograph showing an engineering failure (Figure 9.17A)

Emeritus professor Bob Zulinski, who shared circuit diagrams he had developed in his former life as a teacher (Figure 9.14C)

Associate Professor Ronald K. Gratz, who permitted us to "borrow" from a section he had written on plagiarism (used in Section 3.4)

Associate Professor Tony Rogers and the students in his Consumer Product Manufacturing Enterprise, who gave us access to e-mails and letters they wrote to their corporate sponsor (Figures 4.3 and 4.4)

The reviewers of this manuscript were Rick C. Clifft, Arkansas State University; Suzanne Karberg, Purdue University; Edward G. Dauenheimer, New Jersey Institute of Technology; Roger Friedmann, Kansas State University; Sally B. Palmer, South Dakota School of Mines & Technology; Kirk Schulz, Mississippi State University; Groves Herrick, Maine Maritime Academy; Johann Briffa, Oakland University; Charlotte Brammer, University of Alabama; Carolyn Plumb, University of Washington; and Julia M. Williams, Rose-Hulman Institute of Technology.

Numerous Michigan Tech engineering students during the past 20 years whose difficulties with technical communication have made us laugh and cry

We hope that you enjoy this text. Better yet, we hope that you find it useful in your studies and in your career *and* that you enjoy it.

Sheryl A. Sorby

William M. Bulleit

1

Introduction

1.1 Importance of Communication in Engineering

That engineers are required to have excellent math and science skills is universally understood. What is less well known is that excellent communication skills are nearly as important for success in engineering. Engineers design and manufacture or construct solutions to problems we face as a society. Engineers design devices and systems that improve the quality of our lives. While engineers apply math and science in their design solutions, they also need to be able to *communicate* their design solutions to others so that the designs can be constructed or manufactured. The design solution alone is of little value if no one else understands what is being proposed. Engineers also typically work within corporations or governmental agencies and thus need to be able to communicate with coworkers and managers who may or may not be engineers.

The Accreditation Board for Engineering and Technology (ABET) has recognized the importance of communication skills in their overhaul of accreditation standards for engineering programs. ABET criteria for accredited engineering programs now require graduates to have "an ability to communicate effectively" (Accreditation Board for Engineering and Technology 2003, 2) in addition to understanding and being able to apply math and science to real-life problems. In establishing the current criteria, ABET worked closely with engineers and managers from corporate America as well as engineering faculty to ensure that educational programs are meeting the needs of the companies that will be hiring engineering graduates.

Since nearly all engineers have well-developed math and science skills, also having excellent communication skills will often distinguish an individual among peers. An engineer with excellent communication skills will likely be able to progress through the company hierarchy in ways poor communicators cannot. Achieving excellence in communication skills will enable you to interact with your colleagues more effectively and will earn you the respect of managers and peers alike.

Engineers communicate with one another and with others by several means. Informal drawings and sketches are a predominant method of communication in engineering (e.g., Ferguson 1992) but are not the focus of this text. Instead, this text focuses on many of the other forms of communication in which engineers engage. These forms of communication include written documents, oral presentations, design calculations, and graphical images other than drawings or sketches. In addition, this text contains information about considerations for communication in the electronic age. We hope you will be able to apply the information presented in this text to improve your communication skills, which will enable you to achieve a satisfying and rewarding career in the engineering profession.

1.2 Philosophy and Organization of Text

This book is a comprehensive text that introduces you to the various forms of communication you will engage in regularly as an engineer or an engineering student. The text is a reference you can begin to use during your first year in engineering school and continue to use through a lifetime of effective communication as a student or practicing engineer. For each type of communication we discuss, the format of the presented material is as follows: Function, Form, and Mechanics. For example, when discussing memos in the text, we talk about the general function of a memo, discuss accepted formats for a memo, and then review some of the detailed mechanics involved in writing a memo.

The text covers the gamut of forms of communication engineers might use, including written (e.g., memos, letters, and papers), oral, design calculations, visual, and electronic. A section on audience analysis is included in Chapter 2, and issues on appropriate tool selection are discussed throughout the text since so much of the technical communication in which engineers engage depends on these considerations. Grammar and syntax are discussed in Chapter 2, as well as in the Appendix. Ethical communication, such as the appropriate use of references, is discussed in Chapter 3. Chapter 4 addresses personal communication, such as letters and e-mail. Chapter 5 covers document design, such as font selection and the use of white space. Chapters 6 and 7 are devoted to technical written communication and design calculation communication, respectively. Chapter 8 covers oral communication, and visual communication is discussed in Chapter 9. Chapter 10 focuses on special types of documents you might create when you are seeking employment. The Appendix contains a selection of grammar rules, word usage preferences, and some other considerations.

Each chapter shows you examples of good communication. Some examples of poor communication are also included because we believe that often more learning occurs, for engineers especially, through examples of what not to do than through examples of what should be done. Our examples of poor communication will generally have callouts so that you can clearly see what is wrong with a particular example.

In using this text, you should begin by reading the section on audience analysis in Chapter 2. You should refer to the other sections of Chapter 2 and the Appendix that focus on grammar, word usage, and writing style as you are creating your documents or other forms of technical communication. Chapters 3 through 10 can be studied independently of one another, depending on your needs.

1.3 Communication in the Information Age

In the fifteenth century, the Gutenberg printing press transformed the way books were printed and produced. Gone were the days when monks painstakingly hand copied books. The development of the printing press brought Europe out of the Dark Ages and into the Renaissance and the Age of Reason. Inexpensive books could be rapidly produced, which facilitated the exchange of ideas and information. Universities multiplied, which furthered learning and intellectual thinking.

The microprocessor rapidly became the twentieth century equivalent of the Gutenberg press, and its influence continues to grow. Computers have infiltrated virtually every aspect of our everyday lives. Our cars, our appliances, our homes, our offices, and our leisure activities all now depend on computer systems. We can do our banking and bill paying online, obtain an education online, communicate with one another by modem or other means, send pictures to loved ones, search out new information, and do our shopping from the comfort of our homes using a computer system that is relatively inexpensive, easy to use, and reliable.

Just as the computer has changed our personal lives in untold ways, it has transformed the ways in which engineers communicate with one another and with others. When we (the authors of this text) began our engineering studies, reports were handwritten or occasionally typed, drawings were created by using hand drafting tools, charts and graphs were produced by hand on grid paper made especially for that purpose, and photographs were pasted into the document with rubber cement. Now, all these functions can be accomplished with relative ease through the use of a desktop computer and standard software.

Computers have not necessarily made the process of technical communication easier, but they have enabled us to create professional-looking documents more efficiently and effectively. Engineers must still compose documents by putting words together, including graphics, and following established rules and customs for technical communication. A blank computer screen is just as formidable now as a blank sheet of paper in a typewriter was in the 1980s. Although computers cannot do our writing for us, they have had perhaps the most influence on enabling us to edit and revise our documents relatively painlessly. In the past, a paper would need to be retyped completely if one sentence was to be added or removed. The ease with which documents can be revised and edited through the use of word-processing software has changed our standards for technical communication. A handwritten equation in the middle of a typed document was formerly acceptable; a handwritten equation in the middle of a word-processed document is not acceptable anymore—the word-processing software has built-in capabilities for creating professional-looking equations and you will be expected to use them.

Throughout this text, sections that focus on issues you must consider or understand when you are communicating in this electronic age are included when appropriate. Software instruction is *not* provided. For questions about the use of specific software packages, please refer to the manuals or help windows for the software package you are using.

1.4 **References**

Accreditation Board for Engineering and Technology. 2003. *Criteria for accrediting engineering programs*. Baltimore: Accreditation Board for Engineering and Technology.

Ferguson, E. S. 1992. *Engineering and the mind's eye*. Cambridge, MA: MIT Press.

2
Setting the Stage

2.1 The Writing Process

As engineers, you are likely aware of the steps involved in system or product design. The engineering design process is well established, and you have probably learned about it in some of your courses. Most texts define engineering design as a multistep process, with the number of steps varying according to the author's personal preferences. Even though no consensus on the exact number of steps in the engineering design process exists, experts agree that it consists of multiple steps and that the process is iterative—the first design is rarely the best design.

The *writing process* has much in common with the engineering design process. It is a multiple-step process, and it is iterative. In this section, we discuss the process of writing a document. Experts may disagree about the exact number of steps in the writing process; however, it is a creative process much like engineering design, and, in the end, the result is what matters, not the exact procedure you followed.

2.1.1 The Writing Process Versus the Engineering Design Process

Table 2.1 shows the steps in the engineering design process (adapted from Holtzapple and Reece, 1998) alongside the steps in the writing process. The following discussion of Table 2.1 emphasizes the writing process but may include limited discussion of the design process to show how these two creative processes parallel each other.

1. ***Identify the need and define the problem.*** If no real need exists for a written document or for an engineered solution, you should not waste your time on one. Typically, this step will be done for you in the sense that your instructor, supervisor, or client will tell you what needs to be written or designed. However, in most cases, this step also involves *defining* the need. At this stage, you also define the problem your document is to solve: its objectives. Answer the question "What do I hope to accomplish with this document?" If you need merely to *inform* someone of results you obtained, you will probably use an entirely different tone and style than if you need to *persuade* someone about an action to be taken. At this stage, you will also need to think about whether additional information will be required for preparation of your document, and you may need to perform a library or Internet search to find the documentation you need.

2. ***Perform an audience analysis.*** In the second step, you decide what type of document is appropriate for your audience. (Details of audience analysis appear in Section 2.2.) You must also determine the level or levels of knowledge of your readers that must be addressed in your document. For instance, you may need to write a report that includes a simple summary for a general audience, an executive summary for the company president, and a full formal report for other company engineers, including your direct supervisor. Decisions about audience analysis are analogous to those that must be made when you are designing an engineered system. You must always consider any constraints on the

Table 2.1. The Writing Process and the Design Process

The Writing Process	The Design Process[a]
1. Identify the need and define the problem.	1. Identify the need and define the problem.
2. Perform an audience analysis.	2. Identify the constraints and criteria for success.
3. Plan the document (brainstorming, formal or informal outline).	3. Search for solutions (e.g., brainstorming) and perform a feasibility study and/or develop a preliminary design.
4. Write the document, which usually requires multiple drafts (i.e., an iterative process).	4. Create a detailed design, which is usually an iterative process.
5. Submit the document to the appropriate people.	5. Communicate and construct the solution.
6. Evaluate the results.	6. Verify and evaluate the solution.

[a]Adapted from M. T. Holtzapple and W. D. Reece, *Foundations of Engineering* (Boston: McGraw-Hill, 1998), 436–55.

system design, such as the budget and time limits, as well as design criteria, such as aesthetics, reliability, and maintainability, before the design work can begin.

3. ***Plan the document.*** Careful planning of your technical communication task will greatly increase the efficiency with which you complete the exercise. In the early grades, you were probably taught to outline papers in detail; however, you may or may not have enjoyed that experience. Outlining your documents before you begin to fill in the blanks will help you organize your thoughts so that your document follows a logical order and makes sense to others. You might consider using two basic methods for outlining when you are creating your technical documents. The method you prefer will likely depend on your personal preferences and writing style. Whichever method you choose, make sure you understand that the outline should be a flexible document: as you write and formulate new ideas, you should be willing to abandon your original outline and adopt a new one. Note that at this stage in the writing process, you will likely think more creatively if you use pencil and paper rather than a computer for your document planning. Similarly, at this stage in the engineering design process, using a pencil and paper for sketching often results in more creative solutions than would be obtained with a computer-aided design (CAD) system.

   The first method for creating an outline for a document is similar to the third step in the engineering design process. With this method, you start by brainstorming all the important points you want to make in the document. During your brainstorming session, you simply write down words or phrases as they come to you, in no particular order. After completion of the brainstorming exercise, you then look at all the words and phrases you have written down and start to group them logically according to categories that make sense to you. Once you categorize your topics, you can begin to assign relative importance to each category and start to order the topics for the final document accordingly. After you complete the first draft of your outline in this way, refine it further by looking through it to make sure the order flows from one topic to the next. Move individual topics or entire categories around within the outline as necessary until you are satisfied that following the outline will likely produce the final document you desire. You can then begin to write by expanding on the topics in the outline. You may even want to set up your document with headings and subheadings according to your outline before you begin your documentation task.

   The second method for creating an outline is a bit more structured and may be more appealing to novice writers. In this method, you start with a standard outline for the type of document you are creating and then flesh it out, adding detail according to your specific task. In this text, we supply you with a basic outline for most types of documentation you will be required to create. With this as a starting point, you can complete the outline by filling in details and specific points you want to use. For example, in Chapter 6 you are given the following as the basic outline for an experimental report:

1. Title
2. Abstract or summary
3. Introduction
4. Procedure
5. Basic theory
6. Data and results
7. Discussion
8. Conclusions
9. Recommendations

Using this basic outline as a starting point, you can write beneath each heading the specific points you will be making in that section. For example, under the section titled "Procedure," you might list points like these five:

1. Measured and weighed specimen.
2. Inserted specimen into testing apparatus.
3. Loaded to failure—data sampling at each 500# of load.
4. Made final measurements.
5. *Note:* Include the ID numbers of the testing equipment in report (instructor requirement).

This approach allows you to know exactly what you need to include in each section of the report before you begin your writing task. You might also decide at this point that some of the sections can easily be combined to make your report more readable. For example, you could decide to combine the introduction with the section on basic theory or the conclusions with the recommendations.

When using an outlining method such as this, you will probably need to develop headings that are appropriate to the document you are creating. For example, in Chapter 6, the basic outline for a paper includes a section titled "Main Body." In reality, you would not have a section labeled "Main Body" in your paper. Your main body would consist of the specific headings and subheadings that make sense for the particular topic about which you are writing.

We have emphasized two fairly methodical approaches to planning your document. As you organize and write more documents, you may find that these formal methods are not to your liking. For example, you may find that you just want to write down a few ideas and then dive directly into the writing. Nothing is inherently wrong with planning a document in an informal and personal way. The final result is ultimately what counts. As we pointed out previously, writing the document, like performing the detailed design, is an iterative process, so where you start is a personal decision that can be reached only through practice. As your writing skills mature, you will likely change your processes to suit your personal preferences. You may find that outlining helps you get started and possibly reduces the number of iterations required to progress from nothing to a final document. Alternatively, you may prefer just to start writing and then reorganize after you have words "down on paper." Successful writers work at both these extremes and at all possible combinations between.

Document planning in the writing process is analogous to conducting a solution search and feasibility study or developing a preliminary design in the design process. When searching for a solution to a design problem, an engineer or a group of engineers will consider a broad range of possibilities with only limited criticism of the ideas (i.e., brainstorming). The possible solutions will then be examined more closely through feasibility studies or preliminary design. *Feasibility studies* are rough analyses of the possible solutions to determine which appear workable. *Preliminary design* is a more refined analysis of the possible solutions that look feasible, but it is still only a skeleton design. If only a few possible solutions exist, the feasibility study step might be skipped. From this discussion, you can see that the document-planning step that takes you from nothing to an outline or a preliminary version of your final document is similar to brainstorming and conducting feasibility studies in the engineering design process.

4. ***Write the document.*** Writing the document is the most significant stage of the writing process, and it is analogous to developing the detailed design of an engineered system. In the design process, the engineer or engineers work toward an optimal design by expanding and refining the preliminary design. This stage requires multiple steps, drafts, or iterations that lead the designer to the final result. Often, some steps lead to dead ends or complications that produce significant changes from the preliminary design. As you design, you also need to make full use of codes and standards if they are applicable.

   Writing your final document involves a similar process. From your outline or rough initial version of the document, you must achieve the final document through a series of drafts (i.e., iterations). You must be aware of stylistic requirements appropriate to your task. You should print the document a few times and critically examine factors such as page layouts and white spaces (covered in Chapter 5). Sometimes you may want to set the document aside for a brief period and return to it with a fresh eye. Writing drafts of your document may proceed smoothly or may lead you to rethink the earlier drafts completely, which may require completely reorganizing the document or starting over. Discovering that you must redo work you already produced is unavoidable in both design and writing and is perhaps one of the most difficult facts for novices to accept. Again, the final product is what counts; don't be afraid to abandon solutions or documents that are not turning out as you expected.

One technique you may find useful when you are finished with the first draft of your document, before you revise it, is to create a checklist to guide your efforts. Figure 2.1 shows a sample checklist you might use to guide revisions of your technical document as you work toward your final product.

5. ***Submit the document.*** At this stage in the writing process, you send your document to the appropriate readers and wait for results. You may receive feedback from audience members saying they need more information. You may receive a congratulatory e-mail from your boss. As a student, you will probably receive a grade from your instructor. This stage is analogous to what is typically the largest portion of the engineering design process: communicating and constructing the design. The

---

1. In the initial stages of the process, did you
   ✓ Articulate a clear purpose?
   ✓ Analyze your audience and determine an appropriate style?
   ✓ Brainstorm ideas for inclusion?
2. In creating an outline, did you
   ✓ Follow a standard format?
   ✓ Group elements logically?
   ✓ Order items within a group appropriately?
3. In drafting your work, did you
   ✓ Include references when necessary?
   ✓ Include all necessary information?
   ✓ Think about answering readers' questions?
   ✓ Write in the style you had chosen for your audience?
4. In revising your document, did you
   ✓ Set the document aside for a few days?
   ✓ Read through it carefully and critically?
   ✓ Use checklists such as this for a fresh look?
   ✓ Use the spelling and grammar checker of your word processor?
   ✓ Show it to a colleague for comments?

Figure 2.1. Sample checklist for completing a final document draft

analogy between the writing process and the design process breaks down a little at this point, but the basic concepts of sending out your written document to accomplish its objectives and submitting the detailed design for manufacture or construction are still somewhat similar.

6. ***Evaluate the results.*** When the document has been submitted and either accomplished its objectives or not, you might want to consider how you could have done your writing job better. Possibly the process you used to write the document was not appropriate; for instance, you chose not to outline and the document was sent back to you to be reorganized. Alternatively, maybe the document was well organized, but the recommendations you made cannot be followed. In this case, you must reassess the information and arguments you used to support your recommendations. Even if your document was successful, you should take some time to consider whether it could have been done better. If you are a student, your professor will give you a grade and will probably make comments on your document. Read through these comments to understand how your document met or failed to meet his expectations. Consideration of these factors will enable you to prepare a better document the next time and perhaps even earn a better grade for your efforts. This step in the writing process also holds true for the process of designing engineering solutions. Unanticipated occurrences in the final system should be examined so that you will take them into account in your future designs. Even if the system works as designed, possible methods to streamline the design process, optimize the design, or enhance the manufacturing or construction process should be considered, to guide your future design tasks.

2.1.2 Collaborative Writing

Working in teams is typical of modern engineering practice, and as an engineer, you will often work with others to develop and write documents as part of your total team effort. In this section, we address common issues you may encounter when you are working in a team environment to create technical documents (i.e., *collaborative writing*). In Chapter 8, we examine group oral presentations.

As you know from your engineering design classes, you can adopt certain behaviors that will greatly facilitate collaborative work. These behaviors also apply to collaborative writing situations in which you might participate. Types of behaviors conducive to collaborative work include the following:

- Showing respect for your teammates' opinions by listening carefully
- Defining specific tasks to be accomplished by the group
- Choosing a group leader who will facilitate but not dictate
- Establishing a written agenda for each meeting and following it

■ Defining working procedures and policies for the group (the result is sometimes referred to as a *code of cooperation*) and sticking to them

■ Establishing schedules for task completion and ensuring all members accomplish their goals on time

■ Encouraging participation of every group member by specifying the task or tasks for which each person is responsible

■ Providing constructive critiques and praise when needed

The collaborative writing process requires consideration of some issues above and beyond those discussed previously for individual writing. The overall writing process is the same, but the details of steps in the process require some modification to ensure true collaboration. After the following note, we consider each of the six steps in the process and describe additional considerations for collaborative writing.

Note: Many modern-day word-processing software packages have been optimized for collaborative work. Certain software features allow you to make changes to a document and still leave a version of the previous document visible. In this way, the initial author of the document can view her original work alongside your proposed changes and either accept or reject them. Specific details on how this can be accomplished are beyond the scope of this text; however, you may want to investigate this option with your preferred software package before you begin your collaborative writing task. This option is often referred to as *tracking changes.*

1. ***Identify the need and define the problem.*** In collaborative writing, defining the problem and setting the objectives for your document typically involves a face-to-face meeting among members; however, you may be able to accomplish this task through an e-mail discussion. In this step, the group leader should draft a set of objectives for discussion and allow others to comment or revise them as appropriate until consensus is achieved. Again, for many of your student team-writing assignments, your instructor will likely establish the objectives.

2. ***Perform an audience analysis.*** The audience analysis step in the process is essentially the same as for an individual writing task, except the team members' opinions must be taken into account and a group decision made. Once again, having the team leader draft something for comment is probably the most efficient means for accomplishing this task.

3. ***Plan the document.*** Planning the document is particularly important in collaborative writing because differences of opinion must be settled prior to writing the document. Brainstorming is a group activity and may be more fruitful in collaborative work than it is in an individual effort. Remember to approach brainstorming in a nonjudgmental fashion. Criticizing a team member's ideas during a brainstorming session will likely

short-circuit the entire brainstorming process. At this stage, the group probably needs to work on an outline of the document, and working in a face-to-face meeting rather than through online chats or e-mails is probably best. In a collaborative writing effort, an outline is preferred so that the entire group agrees about the organization of the document, although changes to the outline may still be necessary at some point in the process. A good idea is to have the group agree on format issues, such as the overall document design, fonts, and heading styles, at this initial planning meeting. The group should also select a coordinator to manage the process who will be able to call meetings, enforce deadlines, and generally guide the team through the process. Last, the group should select the deadlines for the various stages of the work.

4. *Write the document.* A group can collaborate to write a document in three basic ways. We discuss them after enumerating them:

> 1. The first method is to divide the document into sections of about equal length and have each team member write a section. Each individual does all his own work on his portion.
>
> 2. The second method requires that one person write the entire document, then pass it on to another team member, who goes through it, makes all the changes she thinks are necessary, then passes it on to another team member, and so on until all team members have had an opportunity to work through it.
>
> 3. The third way is to have one member be a project manager who organizes the task, schedules the work, and, if necessary, manages disputes. A second member then gathers all necessary information, makes notes, and writes a rough draft. A third member, who is a good writer, then writes the entire document on the basis of the rough draft. A fourth member edits the document.

Each method has its strengths and weaknesses. Use of the first method may produce a document with different styles, word usage, and sentence structure in the various sections. The document will need to be made consistent either by one team member or by all the team members in a process similar to that of the second method. The second method will likely produce a document with a single blended style. The potential problem with this method is that the team members must be willing to allow their styles to be blended into a single document. If any team members fight changes to their particular style, team interaction can be hindered, and the document-writing process may be slowed significantly. Fighting changes can also be a problem with the first method if the attempt is made to make the document consistent in either of the ways suggested. The third method is probably the best for industry team writing when an individual can do each step described, but in some situations this method is inappropriate. For instance, a report may have a section on the mechanical engineering for the project, one on the electrical engineering, and one on the materials engineering. Most likely, one team member cannot write all three sections, even if another member has developed a rough first draft. In this case, the first method should be used and combined with the second method to blend the document into a single style.

5. *Submit the document.* Document submission in collaborative writing is the same as for individual work.

6. *Evaluate the results.* The difference between the evaluation step in the individual case and that in the team case is that evaluation of the results should be a team effort. A team discussion of the results of the document submission will help determine the strengths and weaknesses of the document and the process used to write it. Even if the team will never write a document together again, an evaluation discussion will likely help all team members become better writers in future individual and team efforts.

2.2 Audience Analysis

One process you must complete before attempting any technical communication is an audience analysis. *Audience analysis* means you must consider three issues: what your audience members need or want—their *interest,* what they already know—their *knowledge,* and what they know how to do—their *skills.* For instance, if a technical problem arises on a project you are working on, you would need to approach an engineering colleague differently than you would the project manager. If you were required to discuss the problem with the president of the company, you would likely need an entirely different approach to be successful. If you were going to talk to a fifth-grade class about your engineering project, you would need to consider the class's limits carefully. The children's interest level might be high, depending on the project, but their knowledge and skill level would be low. For instance, their knowledge level would require that you use simple mathematics and explain all engineering concepts in simplistic terms they can readily understand. Furthermore, consideration of their skill level would mean, for instance, that they could not read a set of blueprints—something you probably routinely use when you describe engineering projects to other audiences. Deciding on the approach to take in each case is audience analysis, in which the audience is the engineering colleague, the project manager, the president, or the fifth-grade class. Note also that an audience can range from a single individual to a large group.

Audience analysis is a vital component of all technical communication, and whether you are writing or speaking, the first step is to consider the type of audience you will be addressing. The technical writing texts referred to in Section 2.5 discuss audience analysis in various amounts of depth; you should examine a few of them to consider different writers' views. Houp et al. (1998) suggested one possible categorization of audiences that nicely exemplifies the issues. In this categorization, the audiences you could be addressing are executives, experts in your field, technical people outside your field, laypeople, and any combination of the four. In the preceding examples, three of these five possibilities are apparent: your engineering colleague is likely to be an *expert;* the project manager is either an *expert* or an *executive,* depending on the manager's level of engineering knowledge and skills; the president is most likely an *executive;* and the fifth-grade class is a *lay audience.* Let's consider each of these five categories in some detail.

2.2.1 Executives

Executives are primarily interested in the big picture and do not have time to read a large technical report. This fact means they will be interested not in detailed design and analysis information, but in only a brief overview. They will be interested in conclusions, recommendations, and the potential impact on the company, particularly the financial impact, if the recommendations are followed. For instance, an executive will often need to make a decision based on what he reads. Thus, an executive summary is a vital portion of the document when you are writing a technical report for the company. From the standpoint of being able to accomplish something, an excellent executive summary (see Chapter 6) will probably be the most important part of the report.

So, if you were trying to communicate with the company president about a technical problem, you would want to leave out the *details* of the engineering design and analysis and discuss their *implications* instead. The president may be so far removed from "engineering" that she is not interested in the details, and she might not even have the knowledge base or skill levels to understand them. This possibility is particularly true in large companies, where upper-level managers may not have technical backgrounds. Many executives can be considered laypeople when we are thinking about audience analysis, and issues discussed subsequently in the lay audience section of this chapter (Section 2.2.4) also often apply to them.

2.2.2 Experts in Your Field

You should be able to communicate with experts in your field most easily. This audience will generally be interested in the technical details of the project *and* will have the required background knowledge and skills to understand them.

Mathematical expressions, tables, and graphs are appropriate when you are communicating with experts in your field. Technical terms, except those that are extremely specialized, will not need to be defined. For instance, if you are a mechanical engineer discussing your refinement of the internal combustion engine to a group of colleagues, your audience's basic knowledge of mechanical engineering, including terminology, can be assumed, and only specialized terms related to your refinement will need to be defined for this audience.

2.2.3 Technical People Outside Your Field

Technical people outside your field can generally be divided into two types: technical people with an engineering background in a discipline outside your own, and technicians whose training is more hands on and less theoretical than that of a typical engineer. Each group has a technical background but different knowledge and skill levels than yours.

This audience typically has a significantly high level of knowledge and skills, but it will not have the specialized knowledge necessary to understand all terms from your discipline. Definition of terms is most important for this audience. However, your job of communication should be made easier with this group since you can assume a reasonable level of mathematical and scientific knowledge and you can safely use figures, graphs, and tables.

2.2.4 Lay Audience

A *lay audience* consists of people who are far removed from your specialty. They can be difficult to communicate with sometimes because their knowledge and skill levels are generally low with regard to your topic. On the bright side, they are usually interested in the material or they wouldn't be reading your paper or listening to your presentation. The people in a lay audience are not, however, interested in the technical details. They are interested in how things work in a broad sense and in how they will be personally affected by your engineering project. In the case of an executive, he will be interested in how the company will be affected by your project.

The sophistication level exhibited by this type of audience can vary widely. Sometimes, the audience may have reasonable mathematical and scientific backgrounds that will allow a fairly deep discussion of engineering topics. At other times, it will be like the fifth-grade class, and any discussion of an engineering topic will require great care and limited depth. Since a typical lay audience will usually consist of people with a range of abilities, you will not be able to please all the audience all the time. You must define technical terms by using simplistic ideas if necessary, but, when possible, avoid technical terms altogether. Speaking to a lay audience will require that you make special efforts to keep its interest alive. Some ways to do this are to use analogies, to include human interest stories that relate to your topic, and to use pictures and colorful figures. Tables of data and graphs must be used with care since many laypeople will not have the skills necessary to understand fully the implications of your data. When writing for a lay audience, keep your sentences fairly short because long sentences make the material more difficult to understand. In short, try to remember when you knew nothing about the subject, and communicate from that vantage point.

2.2.5 Combined Audience

Effective communication with a combined audience is often the most difficult to achieve because the interests, knowledge, and skills of individuals within the group vary widely. Thus, the best you can hope to do is strike a middle ground. An example of communicating with this type of audience would be if you were asked to speak in front of a group of engineers and their spouses. Again, in this case you will probably not be able to please all the audience all the time, but you must try to strike a balance between providing sufficient detail to hold the engineers' interest and providing sufficient background for their spouses, who may or may not have a technical background. You will need to make a special effort to determine which points you want to get across to the audience and use them as your guide in deciding on an approach. The topics you cover should be selected so that you can engage the most people in the audience.

When considering a combined audience, you may want to separate the audience into a primary audience and a secondary audience. The *primary audience* is the portion of the audience you believe is most important to communicate with. The *secondary audience* is the portion just below the primary audience in importance. Note that this breakdown may not include all audience members. For instance, suppose you are giving a presentation to engineers, their spouses, and some of their small children. You might choose the engineers as the primary audience, choose the spouses as the secondary audience, and leave out the children. The implication of this breakdown is that you want the presentation to appeal mostly to the engineers and to be somewhat interesting and understandable to the spouses; it may or may not be interesting or understandable to the children. You may separate the audience into more than two groups that you will try to communicate with, but typically the audience is divided into two groups at most.

2.2.6 Audience Analysis in Oral Communication

In oral communication, the audience is real, not just an abstract group consisting of "people who will read your report." Consider the following scenario. You must give a 30-minute presentation on a technical topic, going into as much depth as possible given the time and the audience. Imagine that you know the topic well. You must give the presentation to four audiences: a group of fifth graders, a group of high school seniors, a group of college seniors in multiple engineering fields, and a group of engineers who work in the field that relates to your presentation. Can you use the same presentation for each group? Your audience analysis skills are already sufficient for you to answer this question with a resounding *no*.

However, a story about Albert Einstein, probably apocryphal, might suggest otherwise. The story is that Einstein prepared a presentation on physics, thinking it was for kindergartners. When he arrived at the presentation room, he discovered that the audience was physicists, not kindergartners. So, he gave the presentation he had prepared, and it was, according to the audience, the best presentation they had ever heard from Einstein. If this story is true, some other conclusions are possible. Maybe Einstein generally gave poor presentations and trying to make the material understandable to kindergartners helped him give a better presentation for the physicists. Even if he generally gave good presentations, maybe he didn't do a good job making his presentation appropriate for kindergartners, but, in making the effort to do so, developed a better presentation. Of course, if the second scenario were the case, the presentation wouldn't have been appropriate if the audience had been kindergartners. So, even if the story is true, it's not necessarily a reason to argue that you could develop one presentation for the fifth graders and use it for all four audiences. Generally, a presentation on technical material that is suitable for fifth graders would be boring for the other audiences.

So, most likely, to give the presentation to the four audiences, you would need four presentations, one for each audience. Once you would determine this fact, you would need to design each presentation, keeping in mind at all times the audience's capabilities and limitations. If we use the classification of audience types from the previous discussion, the fifth graders and the high school students would each be a lay audience, the college students would be technically competent people outside the area of specialization, and the engineers would be experts. These classifications should help you decide how to design each presentation. Even though the fifth graders and the high school seniors would each be a lay audience, their capabilities would be different. High school seniors would have a larger vocabulary, better math skills, and more life experience than the fifth graders would. This difference would affect the words you could use without explanation, the math you could use, and the general tone of the presentation.

2.2.7 Audience Profile Sheet

As you organize your thoughts about your intended audience, you may find creating an audience profile sheet, such as the one shown in Figure 2.2, helpful. This audience profile sheet could be modified to suit the specific needs of your technical communication task; use of a sheet such as this will help you clarify your thoughts and prepare to begin.

2.2.8 Summary

Audience analysis is a process you must complete whether you are giving a presentation, writing a technical report, writing a paper, or performing a set of design calculations. You must identify the people whom you will be communicating

Primary audience: _____
Secondary audience (if any): _____
Prior knowledge of subject: _____
Level of understanding of jargon: _____
Probable questions: _____
Probable objections: _____
Audience attitude toward subject: _____
Educational level of audience: _____
Information important to this audience: _____
Tone and style: _____
Intended effect: _____

Figure 2.2. **Sample audience profile sheet**

with, decide what their level of background knowledge is, and determine what their needs are. Without going through this process, you may waste both your time and your audience's time.

2.3   Grammar and Style

Grammar and style are based on two more general concepts: syntax and semantics. Basically, *syntax* is the organization of the words in a sentence, and *semantics* is the meaning of the individual words in the sentence. Both are vital to good writing, and both affect the overall meaning of the sentence. In nontechnical communication, having a somewhat unclear sentence is usually not a problem; however, in technical communication, lack of clarity can sometimes yield disastrous results. Consider the following sentence:

The road is thinking about her clouds.

The syntax is correct, but the semantics is wrong. You can read the sentence and it sounds like a sentence, but the meaning is unclear; worse, it's meaningless. Now consider this sentence:

Thinking the her clouds about road is.

The syntax is wrong, and apparently the semantics is also wrong, but it isn't really clear. This example is extreme, but if you aren't careful in your writing, similar but less obvious problems can arise. Last, consider the following sentence:

Susan is thinking about her homework.

Both the syntax and the semantics are correct. The sentence makes sense and has meaning.

From the preceding examples, you should clearly see that when you are constructing sentences, both syntax and semantics are critical and must be carefully considered. The rules of grammar are primarily syntactical and give you the base for building correct sentences. However, you as the writer can choose among a wide range of correct sentence organizations (syntax) as well as a wide range of words to say what you want to say (semantics). Such selection is your job as a writer, and your particular selection determines your style. Clearly, many ways exist to write a correct, clear sentence and many ways exist to write an incorrect, unclear sentence. You must know enough about grammar and word usage to avoid the incorrect sentences and to develop a style that gets your ideas across in an interesting and readable manner.

2.3.1 Grammar

Function. The rules of grammar allow you to compose a sentence correctly. Grammatical rules cover proper sentence structure and the way words change with respect to such considerations as gender, person, tense, and voice. Sentences can then be used to build whatever type of document you are writing. When writing technical documentation, you need to know and follow basic grammatical rules. Poor grammar reflects poorly not only on you as a professional, but also on your company if your documents are distributed outside your organization.

Grammatical rules are fairly fluid within the English language. For example, in the 1970s, ending a sentence with a preposition was considered taboo. Currently, this practice is acceptable in many instances; however, in some cases this practice is still considered unacceptable. The reason this grammatical rule has changed with time is that English grammar rules often follow the spoken idiom, which means that because many people ended sentences with prepositions when they were speaking English, the way they wrote in English gradually changed. As more and more people changed their style of writing English, ending sentences with prepositions became more commonplace and hence acceptable. Changes like the one just described should not be taken lightly; you as a writer should, in most technical writing, err on the side of tradition. The former way is usually considered better until a significant number of other writers have changed to the new way.

Form and Mechanics. Selected grammatical rules are discussed in the Appendix. The Appendix does not present an exhaustive discussion of grammar since entire books have been written on the topic. Rather, it is a selection of guidelines based on difficulties we have seen since the 1980s while reading engineering students' writing. Two easy-to-use resources are highly recommended for those of you who want to become better writers. These resources are concise, are readable, and will take you beyond the grammar covered in this text. The first is an updated version of a short, classic writing reference—*The Elements of Style,* by Strunk and White (2000)—which should always be on your desk for quick answers to grammatical questions. The second is a more recent, and slightly longer, book that approaches grammar in a light and usable way: *Sleeping Dogs Don't Lay,* by Lederer and Dowis (1999). As you write more, you will likely require some more formal technical writing references such as Alred, Brusaw, and Oliu (2003) or Hargis et al. (2004).

2.3.2 Style

Function. *Style* is the individual way you write. It sets your writing apart from others' writing. The only way to develop your personal style is to practice writing regularly. The style you will use in writing will vary greatly depending on your audience. For example, if you are writing a letter to your parents, including incomplete sentences, using abbreviations, and fracturing all grammatical rules if you so choose are perfectly acceptable. If, however, you are writing a letter to a client, your style will necessarily be more formal and reserved. In a business letter, you should choose your words carefully so that your meaning is clearly conveyed. Remember that many times technical documents may be used to establish a paper trail through the design process. You need to make sure your letters accurately reflect your intentions; otherwise, your words may be used against you someday. If legal proceedings ever occur for one of your engineering projects (and you should assume that such a scenario is always possible), your choice of words and writing style may become critically important.

Style is not format. *Format* is the manner in which you lay out your document. For instance, we subsequently discuss the *form* of a technical paper—in other words, the *format* for a technical paper. Simply stated, style is the way you write, and format is the way your document looks.

Form. Generally, style is not based on rules but on your personal approach to constructing sentences. This approach requires a good grounding in grammar and much practice. However, in technical communication, you are subject to some limitations on the range of decisions you can make in your approach to style. The stylistic limitations are primarily related to the discourse community in which you are working. A *discourse community* is simply a group that through the years has developed a common way or style of communicating. For instance, engineers often use passive voice in technical documents; however, writers in other fields may prefer active voice. (Passive voice and active voice are described subsequently.) You are probably aware of documents prepared by lawyers that are written in a style often referred to as *legalese*. This style is the accepted format for writing within the legal profession. Many discourse communities exist, so you should learn to imitate the preferred style in your profession to communicate with your colleagues effectively.

Any organization you work for will likely have its own guidelines and formats for writing both internal and external documents. The company itself is a discourse community, but it is also part of a larger discourse community. Thus, the company guidelines for internal documents may be different from its guidelines for external communications. Clearly, one aspect of audience analysis and document planning is consideration of the discourse community. Often, the impact of the discourse community on your writing is implicit in the sense that you do not actually decide which one you are operating in; you simply adopt its style, just as a fish doesn't think about the water it breathes. Other times, particularly if you are writing for a discourse community with which you are unfamiliar, you will need to explicitly modify your style to suit that community. Your writing style may take three basic forms in technical communication.

Informal style. Sometimes when writing a technical communication, you can use a relatively informal style. This style is typically reserved for communications with your peers at your company. An informal style is also appropriate when you are writing a document for use by the general public (i.e., people with no technical background). In an informal style, using the first person to convey your ideas is perfectly acceptable. For example, you could use a sentence such as the following:

> We believe the use of concrete instead of steel is appropriate for this particular project.

Semiformal style. A semiformal style should generally be used for most types of technical writing, especially when your document will be distributed outside the organization with which you are involved. With this style, you should avoid the use of first person unless you are certain such use is acceptable: some technically oriented people consider the use of first person unprofessional. As an example of the semiformal style, the previous sample sentence could be rewritten as follows:

> For this particular project, the use of concrete instead of steel is appropriate.

Formal style. The formal style is most suitable for writing that will be published in technical books and journals. Generally, for this type of technical writing, the passive voice should be used and first person should be avoided. Again, exceptions exist. You should follow the requirements of the publisher to whom the paper will be submitted. The sample sentence rewritten in the formal style follows:

> After careful consideration of several factors, it was determined that concrete rather than steel would be the appropriate material for this project.

If you look through this text, you will find that three distinct styles were used in writing it. First, the Preface was written in an informal style and includes several fractured grammatical rules. We tried to use a humorous, conversational tone because, to us, the authors, the Preface is our chance to have a conversation with you, the reader. Second, the text was written in a semiformal style. We liberally used first and second person throughout; however, we tried to follow all the grammatical rules and to write in a serious tone for the most part. This semiformal style is the accepted writing style for textbooks. In the 1970s, texts were written much more formally, but publishers found that students did not respond well to the formality, so the style was relaxed to improve the readability of the texts. Third, the example documents we included throughout the text were all written in a formal writing style. Since most engineers involved in technical communication still adhere to a formal writing style, we used this style when writing our examples. In some cases, you may be able to write your technical documents in a semiformal style; however, when in doubt, err on the conservative side and write in a formal style.

Mechanics. The following general rules for writing style should be followed, when possible, in creating technical documents.

   First person. *First person* involves the use of words such as *I, me, we,* or *us.* First person is generally avoided in technical documentation, except as discussed previously. A current trend is toward the use of first person in technical writing, but such use is still unacceptable in many instances. Ask your instructor or supervisor about her view on this issue.

   Passive versus active voice. Writing something like "We weighed 50 grams of compound A and put it in a beaker" is referred to as the *active voice.* In the active voice, the subject, *We, performs* the action, *weighed.* Conversely, in technical communication, writing "Fifty grams of compound A were weighed and placed in a beaker" is usually preferable. This style is referred to as the *passive voice.* In the passive voice, the subject, *Fifty grams, receives* the action, *were weighed.* Note that who (or what) is performing the action is not apparent in the passive voice. Most composition instructors frown on the use of the passive voice, and some grammar checkers will even alert you to the presence of the passive voice and suggest that you change sentences from the passive voice to the active voice. For most technical reports, however, the passive voice is still preferred, although this practice may be changing. Note that you may sometimes need to use the active voice for clarity. For instance, instructions or procedures are better in the active voice. The instruction "The wheel is turned 90 degrees" is in the passive voice. "Turn the wheel 90 degrees" is in the active voice and sounds better. To spot the passive voice in your writing, look for *are, is, was,* and *were,* which are the forms of the verb *to be* that are most often used in sentences written in the passive voice. For example, "The phase diagram is shown in Figure 5" is in the passive voice, but "Figure 5 shows the phase diagram" is in the active voice.

   The passive voice is often used in technical writing because the use of first person combined with the active voice seems to emphasize the individual or individuals who did the work rather than the work itself, which seems somewhat subjective. Consider the following sentence:

Active: We analyzed the data using a fast Fourier transform.

This sentence structure seems to emphasize the people who did the analysis rather than the analysis. Now consider this sentence:

Passive: The data were analyzed by using a fast Fourier transform.

In this sentence, the emphasis is on the analysis. Note that if this were an instruction, you could write the following:

Active: Analyze the data using a fast Fourier transform.

A current trend is toward use of the active voice, including the use of first person, in technical writing, but when in doubt, use the passive voice. At a minimum, you should be aware of the two voices and try to use the one that is most effective. Always check with your supervisor, your publisher, or, in the case of students, your instructor, about his or her preferences.

Conciseness. The quality of a report is much more important than its quantity. Make sure your document includes all the necessary information without being repetitive or belaboring any points.

Sentence length. Make your sentences as concise as possible. Generally, in technical writing, sentences should not exceed about 15 words. However, this rule is not hard and fast since you want a variety of sentence lengths in your document for better readability. Too many short sentences make your writing sound choppy. Sentences that are too long can be difficult for the reader to comprehend. Long, flowery sentences are wonderful in fiction but can be a distraction in technical writing. An occasional long sentence is acceptable, but make a conscious effort to restrict most sentences to less than about 15 words.

Bottom line up front (BLUF). In most technical writing, you should state your significant conclusions, findings, results, and recommendations at the beginning of your document. This practice is sometimes referred to as *bottom line up front,* or *BLUF.* BLUF should be used in nearly all types of technical writing, including memos, letters, reports, papers, and any other type of document you may be required to produce. By putting the bottom line up front, you set the stage for the reader and let him know the purpose of the document immediately. Setting the stage for your reader lets him know what is coming and where you are going. With the conclusions in mind, a reader can think about how the results support the conclusions as he is reading the section on results. Also, you want to make sure your most important findings are clearly articulated. If the significant results are buried somewhere on page 30 of the document, the reader will most likely not recognize the results or their significance when he gets to that point in the document. Your boss and your college professors want to know primarily what you discovered. They may or may not be interested in learning how you arrived at your conclusions, but they can read the rest of the document if they are interested in those details.

Sexist language. Sexist language favors one sex over the other, generally men. In the past, the use of male-gender pronouns (e.g., *he* and *his*) was considered correct no matter which sex was being referred to. Furthermore, words such as *chairman, fireman, workman,* and *stewardess* were standard for general reference even though each is gender specific. The current trend is to find ways not to be gender specific. So, for instance, the word *chairman* should be replaced with *chair* or *chairperson*; we recommend *chair.* The word *fireman* should be replaced with *firefighter, workman* should be replaced with *worker,* and *stewardess* becomes *flight attendant.*

Removal of pronouns such as *he* and *his* is often more difficult. The first way to do so is to use *he or she, he/she, his or her,* or sometimes *s/he.* All these constructions are awkward, so they should be used sparingly. An example taken from this text is as follows:

> Always check with your supervisor, your publisher, or, in the case of students, your instructor about his or her preferences.

A second way is to speak to the reader directly by using *you* or *your:*

> You should look at the previous example for the way to use *your.*

A third way is to use the plural. Consider the next sentence:

> An engineer should always check his work.

This sentence uses the gender-specific word *his.* Using the plural of *engineer* produces the following:

> Engineers should always check their work.

Now the sentence is gender neutral. An *incorrect* way to make the sentence gender neutral would be to write this:

> An engineer should always check their work.

Engineer is singular, and the referential pronoun *their* is plural.

A fourth way to remove gender-specific pronouns is to change the verb form. The following sentence uses the gender-specific pronoun *he:*

> An engineer must pass the PE examination before he can become a registered engineer.

Changing the form of the verb *become* yields the following:

> An engineer must pass the PE examination before becoming a registered engineer.

Table 2.2. Fancy Words Versus Plain Words

Fancy Word	Plain Word
commence	start
employ	use
endeavor	try
finalize	end
initiate	begin
prioritize	rank
procure	get
terminate	end
utilize	use

The fifth, and last, way is to alternate *he* and *she* from one paragraph or section to the next. This technique was used in this book, so you should have seen some examples of it. Some of you were probably taken aback when we referred to an engineering manager as *she*. This type of response is a strength of this method. Alternating the pronouns *he* and *she* points out how subconscious some of our gender assumptions are, especially when *she* is used in a nontraditional context.

Plain and fancy words. Use plain words instead of fancy words unless you have a good reason not to. A saying that illustrates this rule is "Don't utilize *utilize* when you can use *use*." One potentially good reason *to* use a fancy word is if you have been using the plain word too many times and need to add variety to your writing. Even better would be to rewrite the offending paragraph or sentence so that you can use the plain word less and not use the fancy word. Fancy words have their place, but use them sparingly in technical writing. Table 2.2 shows some examples.

2.4 Exercises

1. Imagine giving an oral presentation about a design project you worked on in one of your engineering classes. Fill out and turn in a profile sheet for each of the following individuals or groups that might be a potential audience for your presentation:

   a. Your engineering instructor
   b. Your parents
   c. Your classmates in the course

2. View a corporate Web site and identify elements on the page that are geared toward a general public audience and elements that are geared toward a person with a technical background. Print the page and discuss these elements in a memo to turn in to your instructor.

3. Form a small group and select two articles on the same topic. One article should be from a magazine geared toward the general public and the other should be from a technical journal. Working alone, analyze differences in tone, organization, types of graphics, sentence length, and writing style. Convene a group meeting and compare and contrast your analyses. As a group, write a short paragraph, to turn in to your instructor, outlining your team's findings. Within this paragraph, include similarities and differences in individual assessments within the group.

4. Find two advertisements for automobiles, either on the Web or in magazines. One should be for an economy car and the other for a luxury car. Write a paragraph, to turn in to your instructor, about differences between the ads and who you think was targeted for each in terms of age, gender, socioeconomic status, and so forth.

5. Choose a 100- to 200-word abstract from an article in a professional journal. Rewrite the abstract for a nontechnical audience. Turn in to your instructor both the original and the revised abstract.

6. Referring to Exercise 1, create a brief brainstorming list of the style considerations you would keep in mind when writing for the three listed audiences.

7. Choose any product currently on the market that you think has problems from an environmental standpoint. Create a brief brainstorming list of topics that would be included in a report about the product.

8. Choose any product currently on the market that you think has problems from a functionality standpoint. Create an outline of a report that you could write to convey your observations to someone else.

9. On the basis of writing you have done for other courses, create a checklist you can use for future document preparation. Be sure to include items on your checklist that target your known problems in communication.

10. When you are assigned to a collaborative writing group, meet with the group to brainstorm a code of cooperation for collaborative efforts. As a group, organize and revise the code by e-mail until consensus is achieved.

11. Most word processors have built-in outlining capabilities. Modify the six-step writing process discussed in this chapter so that it has nine steps, by breaking some of the steps into substeps. Use the outlining capabilities of your word processor to outline the steps in your newly developed writing procedure.

12. Write a 300-word paragraph describing what you think is the most significant problem on your campus. Use the BLUF technique when writing this document.

13. Do Exercise 12 as a collaborative writing exercise with at least two other people.
14. Do Exercise 12 but write the entire paragraph in the passive voice. Then rewrite it in the active voice. Discuss the strengths and weaknesses of each.

2.5 References

Alred, G. J., C. T. Brusaw, and W. E. Oliu. 2003. *Handbook of technical writing.* 7th ed. New York: St. Martin's Press.

Hargis, G., M. Carey, A. K. Hernandez, P. Hughes, D. Longo, S. Rouiller, and E. Wilde. 2004. *Developing quality technical information: A handbook for writers and editors.* 2nd ed. Upper Saddle River, NJ: Prentice Hall.

Holtzapple, M. T., and W. D. Reece. 1998. *Foundations of engineering.* Boston: McGraw-Hill.

Houp, K. W., T. E. Pearsall, E. Tebeaux, S. Cody, A. Boyd, and F. Sarris. 1998. *Reporting technical information.* 2nd ed. Scarborough, Ontario: Allyn & Bacon Canada.

Lederer, R., and R. Dowis. 1999. *Sleeping dogs don't lay.* New York: St. Martin's Press.

Strunk, W., Jr., and E. B. White. 2000. *The elements of style.* 4th ed. Boston: Allyn & Bacon.

3

Ethical Issues in Technical Communication

The National Society of Professional Engineers has developed a code of ethics for the engineering profession (www.nspe.org). Although not all engineers are Professional Engineers, most feel an obligation to follow an ethical code of behavior in the conduct of their jobs. Ethical behavior for practicing engineers is especially important since many of the products and processes engineers develop have a large impact on human life and the environment. As with all other aspects of engineering practice, technical communication must be developed according to an accepted norm for ethical behavior. Ethics is typically a significant factor in three main areas in technical communication: reporting all relevant information accurately; recognizing legal and social ramifications of personal communication, especially with respect to stakeholder groups; and giving credit to others when appropriate. Each of these three main areas is discussed in a subsequent section of this chapter. Another key area of ethics in technical communication is plagiarism, which is also discussed in a separate section of this chapter.

3.1 Reporting All Relevant Information Accurately

As an engineer, you will likely find that your written work will be subject to close scrutiny, especially in the case of litigation. Since your reputation as a professional will be on the line, you should make sure all your documents meet the highest ethical and professional standards. For this reason, you must include all information accurately and in its entirety.

One situation that frequently arises in engineering practice is reporting experimental data. When reporting such data, make sure you include *all* the data—even data that do not agree with your findings or conclusions. Experimental work usually results in a few data points that are *outliers*. In some cases, outliers are not included in statistical calculations of quantities such as means or standard deviations. If you ignore the outliers in your data analysis, indicate this fact in your report and discuss the criteria you used to decide which outliers to ignore. Include the outlying data in the data tables so that readers can clearly see that you have been comprehensive in your studies and that you are not trying to ignore data that do not fit your conclusions. Table 3.1 shows a typical way to include outlying data in your report.

In many cases, you will be expected to include data and an analysis of the data in your technical documents. Some of the key methods for analyzing data involve the use of statistics. As an engineering student, you will likely take a course in statistics at some point during your studies so that you can analyze data accurately. When reporting the results of a statistical analysis, you will usually report the p value and then state whether this p value indicates that the results of your statistical test are *highly significant* or perhaps *marginally significant*. The acceptable p values for statistical significance will vary by field: what is regarded as highly significant in one field might be considered marginally significant in another field. If you are preparing a document that includes the results of statistical analysis (i.e., p values), you should learn what is considered highly or marginally significant for that field. A good rule of thumb is that if $p < .01$ (this means the probability is greater than 99%), the data are highly significant, but if $p < .1$

Table 3.1. Indicating Outliers Within a Data Table

Applied Voltage (volts)	Current (amperes)
1.5	0.11
3.0	0.12[a]
4.5	0.35
6.0	0.50
7.5	0.61
9.0	1.93[a]
10.5	0.81
12.0	0.92
13.5	1.02

[a] Indicates outlying data not included in analysis.

(i.e., the probability is greater than 90%), the data are marginally significant. For *p* values between those two values, you should use your best judgement.

The preceding discussion is by necessity limited. To perform statistical analyses such as those mentioned, you will need to consult a book on statistical methods to learn more about such topics as types of statistical tests, null and alternative hypotheses, level of significance (indicated by the *p* value), level of confidence, and other information that is far beyond the scope of this text. See, for instance, Walpole et al. (2002).

Complete and accurate reporting is also required in the documentation of numerical studies. When conducting numerical studies, you will typically make several assumptions when applying boundary conditions and initial conditions. For example, you may assume that certain locations are fixed in space or that certain temperatures exist at given points in the numerical models. The results you obtain are highly dependent on the boundary and initial conditions you supply. If anyone else were to try to verify your findings later, he would need to know the assumptions you applied before he could duplicate your results. If another person wanted to extend your work, knowing the boundary and initial conditions you applied would enable her to pick up where you left off easily and seamlessly. Most important, if litigation ever resulted from your project, you would be required to explain and justify all the assumptions that went into your analysis. If you did not accurately record your assumptions, making your case successfully in court would be nearly impossible.

3.2 Adhering to Ethics in Personal Communication

If you are working for a company on a project that involves a product or process that generates revenue, all documents you create during the design and testing phases of the project will likely be owned by the company, or *proprietary*. You may also be working on proprietary projects as a student if you have been contracted through a corporation for a research or design project. All reports and design documents related to proprietary projects must be kept strictly confidential and can be obtained by others only through legal action. Since the company owns the results from proprietary projects, you are not free to distribute the results at will. As an engineer or as a student working on a proprietary project, you are ethically responsible for not revealing information about the project to other individuals. Revealing corporate secrets is strictly prohibited and will often result in dismissal or, worse, litigation.

In contrast to proprietary documents, nonproprietary documents can be freely obtained by invoking the *Freedom of Information Act (FOIA)*. The FOIA is a federal law that enables you to have access to items like your personnel file so that you are fully aware of what your supervisor has included in it. If you are working in a university or another public institution, almost all documents you produce while working for that institution are subject to the FOIA. Items that can be obtained include memos, letters, and e-mails.

Most companies and universities also have policies about what can and cannot be sent through their e-mail system. You should familiarize yourself with these policies and follow them strictly. Policies typically cover considerations such as whether non-work-related e-mails are permissible or whether you can send e-mails off-site. Almost all companies and universities have policies prohibiting sending e-mails with pornography attached or e-mails that could be considered harassing. Be aware that all companies and universities have access to your e-mail accounts and can monitor them as they see fit. Failure to comply with e-mail policies will often result in disciplinary action and could result in job termination or expulsion from school.

If your work is a public works project, common in civil and environmental engineering, you must disclose all relevant information to the citizen stakeholders who will be affected by the project so that they can make an informed decision about whether they want to proceed. For example, if one potential route for construction of a road will mean the destruction of several acres of wetland habitation and another potential route will mean a decrease in property values for some residents, the citizens have a right to know these issues before a final decision is made on routing. Issues related to your ethical requirements when your client's needs do not intersect with the public needs are potentially complex and beyond the scope of this text, but be aware that how you proceed with communication of information may not be easy to determine in some situations.

3.3 Obtaining Information from Other Sources and Using It Ethically

Two main sources of information are available to you for inclusion in your technical documents: written materials in the form of books and journals, and documents posted to the World Wide Web. To prepare technical documentation that is relevant and complete, you must be able to find the information you need effectively and efficiently. Once you find the needed information, your ethical responsibility is to cite the source of the information and give credit to its original author or authors.

3.3.1 Conducting Library Searches

Function. Public and university libraries are typically the most reliable sources for written documents. As such, they are excellent sources for obtaining background information you may need to prepare your technical documents.

Form. Technical (nonfiction) books in the library are organized according to the Dewey decimal classification system. The Dewey decimal system was conceived in the late nineteenth century and has been accepted in most parts of the world.

With this classification system, knowledge (in the form of books) is arranged by subject matter. The 10 main classifications are as follows:

000	Computer, information, and general reference
100	Philosophy and psychology
200	Religion
300	Social sciences
400	Language
500	Science
600	Technology
700	Arts and recreation
800	Literature
900	History and geography

The first digit in the classification system represents the main classification, the second digit indicates the division, and the third digit represents the section. For example, a call number of 532 represents science as the main class, physics (3) as the division, and fluid mechanics (2) as the section. Thus, if you wanted to find books about fluid mechanics, you could browse in the section of the library with books having call numbers in the 532 range. After the first three digits of the classification, a decimal point is added and additional digits are included to specialize the call number further. For example, a call number of 624.23 is used for suspension and cable-stayed bridges, 370.15 is used for educational psychology, 624.0285 is used for computer applications in civil engineering, and so on.

Most university libraries have, in addition to books, collections of journals and magazines. The information in journals and magazines is typically more up to date than that in books and thus represents a significant resource for you as you search for background information on a topic. Journal articles typically undergo *peer review* before publication. What this means is that articles submitted to the journal are sent to experts in the given field who read them and comment on the validity of the results presented. In contrast, magazine articles are typically not peer reviewed (although in some cases, particularly in Institute of Electrical and Electronic Engineers [IEEE] publications, they are peer reviewed) and sometimes may even represent articles solicited by corporate sponsors. As a result of the peer-review process, journal articles are generally considered to be more objective than are magazine articles (i.e., their results are often considered to be more reliable).

Journals and magazines are also typically registered in an online database. Most libraries have developed means for finding articles through online database searches and have written instructions or training sessions aimed at helping you conduct these searches effectively. If you need help with this kind of search, contact a staff member in your library, who will likely be more than willing to help.

Mechanics. Nearly all libraries have online search engines you can use to locate the information you seek. Books can typically be located by searching for an author, a title, or a subject. When searching by subject, you should be careful to include appropriate keywords or phrases. You can also often browse by call number. Thus, if you know the specific call number for the topic you are researching, you can electronically browse the library holdings of all books within the same range of call numbers. As an engineering student, you should learn to use your library resources effectively. Most librarians will be eager to assist you in finding the information you need and willing to help you familiarize yourself with the specifics of conducting searches in your library.

3.3.2 Conducting Web Searches

The World Wide Web, or just the Web, began in the late 1980s when physicists in labs around the world desired a means for sharing data electronically and effortlessly. In the beginning, a person needed to know exact Web addresses to find documents or posted information. As the use of the Web increased, search engines were required so that any user could easily find the information he sought. Currently, the Web is many people's single most important source of information. Because the Web has become an important source of information, you need to learn to use this resource effectively. Information on the Web is communicated through Web sites. Typically, Web sites are linked to one another so that if you need more information on a topic, you can click a link and be immediately redirected to a new Web site. Since all sites are interconnected and interlinked, the overall structure is analogous to a spider's web, hence its name.

Function. Web searches are typically conducted for one of two reasons: to learn all there is to know about a topic or to find one specific piece of information you know is out there but are not sure just where. Since a great deal of information is on the Web, being able to wade through it all is a key to the successful use of this resource. The following Web site is an excellent resource for conducting online searches:

http://www.sc.edu/beaufort/library/pages/bones/boness.html

Form. A Web site is designated by its URL (universal resource locator). Most URLs include the following parts:

http: hypertext transfer protocol, the format used to transfer
        information

www   World Wide Web

In addition to these parts of a URL are a domain name for the location of the server on which the site resides and a top-level domain name that indicates the type of site the server has been designated. For example, Michigan Technological University's Web site has a domain name of *mtu.edu*. In this case, *mtu* signifies the domain name for the Michigan Tech server, and *edu* signifies that this site is owned by an educational institution. Only a few top-level domain names are recognized; however, this situation will likely change rapidly in the coming years as the Web expands in almost exponential growth. The most commonly used top-level domain names are as follows:

.edu   educational site (usually a university or a college)

.com   commercial or business site

.gov   U.S. federal government site (nonmilitary)

.mil   U.S. military site

.net   networks, Internet service provider, or organizations site

.org   nonprofit organization site

Other top-level domains that are also currently available but not used as widely are the following:

.aero   sites restricted by use for the air transportation industry

.biz   sites for general use by businesses

.info   sites for general use by both commercial and nonprofit organizations

.name   sites for general use by individuals

.museum   sites restricted for use by museums

.coop   sites restricted for use by cooperatives

.pro   sites restricted for use by certified professionals

Since the Web began in the United States, a country code was not originally assigned for sites located on servers in the United States. For sites that exist in non-U.S. countries, the two-letter country code for the host country is also included in the domain name. For example, *curtin.edu.au* is the domain name for Curtin University of Technology in Perth, Australia. Other common country codes are .uk (United Kingdom), .de (Deutschland, or Germany), .fr (France), and so on.

Just as you must carefully distinguish between information in journal articles and that obtained from magazines, you should also take the same care in evaluating information you find on the Web. You must evaluate information gleaned from the Web to determine whether it is of value. Since virtually anyone can put anything on a server and create a Web site, all kinds of sites do not contain relevant or useful information. Several sites have also been created as hoaxes, so you should carefully examine information you glean from the Web before you insert it into a research paper or a proposal to your boss. You can look at a few simple factors to help you judge the value of the information on the site:

■ Look at whether the Web page has an author and the date of its last update listed. Sites that do not contain these basic pieces of information could be hoaxes.

- Check for the name of a contact person who will answer questions about the site. Web pages that do not have a way for you to contact someone about your questions are probably not legitimate.
- Be skeptical of pages that have .net, .com, or .org as their top-level domain name. These sites can be created by virtually anyone, with no oversight or limits.
- Sites that have .edu or .gov in their URL are more likely to be legitimate than those with .com. Remember that .com Web sites exist to help a business promote the corporation, so information may be given the corporate spin on a .com Web site.
- Double-check the information presented on a given page by seeing whether it is corroborated at another site. Also check the pages to which a given site links. If the site seems to be linked to several other questionable sites, chances are, the original site is a hoax.

If you do check out a Web site and believe it is legitimate and you ultimately use the information you find there in a paper or report, be sure to cite the reference in your document (just as you would for a paper or a book). Proper citation techniques for Web pages are found in Section 3.3.3.

Mechanics. Several search engines are available for conducting online searches. The one you use will often depend on your personal preferences and on availability. Subtle differences exist among search engines, but most operate similarly. Before you begin your online search, you should first *think about what you are trying to achieve.* Just as you wouldn't go into a library and merely start wandering among the stacks of books, you shouldn't start your online search without some upfront planning.

How you conduct your search will likely depend on what you hope to achieve. If you are trying to locate a specific piece of information, you will want to focus your search narrowly so that the one document you want is one of the few you might find. If you want to learn all that is available on a certain topic, you will want to focus your search with keywords and phrases but not narrow the search too much. If you just want to poke around and have time to kill, you will probably want to search broadly and see where your search takes you.

Search engines work by scanning through the millions of documents on the Web and finding words or phrases you input. If you input the word *ventilation* into the search field, your search engine will return a list of all documents on the Web that contain that word. Probably millions of online documents contain the word *ventilation,* so the documents will be prioritized by the search engine according to the number of times the word appears in a given document: documents that contain the word 50 times will be ranked higher than those that contain the word only once. You can look through the list of documents in which this word was found, click on a link to the page, and then peruse the document to determine whether it is useful to you. Search engines may be *case sensitive,* which means that if you input the word as *Ventilation* instead, the search engine will list only the documents that contain the word starting with an uppercase letter.

If you insert more than one word in the search field, the search engine will typically return any document that has any of the words you input. (In some cases, default settings of a search engine will allow the engine to return only the documents that contain all the words.) Documents that have all the words you input will be ranked higher than those that contain only one or two of the words.

If you are looking for a particular phrase, you should enclose the entire phrase in quotation marks. For example, if your search field is *"sustainable development,"* the search engine will return only the documents that contain that exact phrase. If the search field were merely *sustainable development,* the search engine would return all documents with the word *sustainable* in them and all documents with the word *development* in them. In this case, the documents with the phrase *sustainable development* in them would be ranked higher in the list.

You can use plus signs (+) and minus signs (−) to narrow your search further. For example, if your search field is *+ventilation −heating,* documents that contain the word *ventilation* but do not contain the word *heating* will be returned. You can also conduct Boolean searches using the operators *AND* and *OR* (these operators must be typed all uppercase). In this case, you could search for *ventilation AND heating* (only documents that contained both words would be returned) or you could search for *ventilation OR heating* (documents that contained either word would be returned).

Besides the basic Boolean operators of *AND* and *OR,* two other operators can often be used in performing Web searches: *AND NOT* (or sometimes just *NOT*) and *NEAR.* Boolean operators can also be "nested" within a search field. In this case, you put certain Booleans within parentheses to indicate they should be performed first (similar to the rules of algebra). Table 3.2 includes examples of Boolean searches and their results.

Most search engines have a list of *stop words* that are internally defined. These words are typically the little words (*and, in, of, at, to,* etc.), and they are ignored in the search. In most cases, stop words are ignored even if they are within phrases. For example, a search on *"to be or not to be"* might yield no results because all the words in the phrase have been ignored by the search engine. You might want to check the settings on a particular search engine before you try using it for some searches.

You can also use wild cards (*) to help you find an exhaustive list when more than one form of a word might interest you. For example, if you search for *heat*,* your search engine will return documents with *heating, heated,* or *heater* in them. The wild cards can also be used within phrases and within Boolean operations in a search field.

Table 3.2. Examples of Boolean Operators and Search Results

Search for	Results
(heating AND ventilation) OR "air conditioning"	Documents that contain either the phrase *air conditioning* or that contain both *heating* and *ventilation*.
"air conditioning" AND NOT ventilation	Documents that contain the phrase *air conditioning* but only if they do not contain the word *ventilation*.
(heating AND ventilation) OR ("air conditioning" AND cooling)	Documents that contain either *heating* and *ventilation* or those that contain *air conditioning* and *cooling*.
heating NEAR ventilation	Documents in which the words *heating* and *ventilation* are found within a certain proximity to each other. The proximity is usually predefined for a given search engine but may be adjustable through various settings.

3.3.3 Citing References

In almost all forms of technical writing, you will need to include background or related information that is not your original work. The most common occurrences of doing so are citing references, such as books, journals, or Web sources, and including equations in your document. These topics are covered in the following sections of this chapter.

Function. In most technical writing, you must credit other authors for their ideas by citing their work. In fact, using someone else's ideas or copying sections of text and using them in a proposal, a report, or a paper without giving proper credit is highly unethical. In general, a document can include figures and equations from other sources as long as you cite the reference. If the document will be published by a for-profit organization, however, you will need to obtain the copyright owner's permission before you can include items such as figures in your document, even with proper citation.

Form. You should list all the papers and books cited in your paper or report at the end, in the references or bibliography section. The way you cite references and include them in reference lists typically follows one of three accepted styles: APA, from the American Psychological Association (2001); CBE, from the Council of Biology Editors (Style Manual Committee 1994); or MLA, from the Modern

Language Association (Gibaldi 2003). The citation method you choose will depend largely on your audience. Many engineering audiences prefer the CBE method of citation; however, if you are preparing a report for a specific technical journal or for an instructor in your English department, you may need to use either the APA or the MLA style. The similarities and differences among these three citation styles are discussed next.

With CBE citations, you typically cite references in the order in which they appear in the document. Citations are either superscript numbers or numbers contained within a set of parentheses. The first reference is [1], the second [2], and so forth. If you are citing more than one reference at a given location in your document, you include the numbers as either a list, if they are not sequential, or a sequence (e.g., 1, 3, 9 or 2–4). If you use a reference a second time, you refer to it with its original number; you do not assign it a second number. In the reference list at the end of the document, you consecutively list the items by number, making sure the number used to identify a reference in the text corresponds to its number in the list.

An alternative CBE citation method is to cite the references by including the author's name and the publication year in parentheses within the document text. A correct citation in this format is (Duits 1993) for one author, (Duits and Smith 1994) for two authors, and (Duits and others 1995) for more than two authors. If you want to include the author's name in the sentence, the citation would look like this: "A study by Duits (1993) showed that. . . ." If this citation method is selected, references are listed in alphabetical order by the last name of the first author of a given paper or book.

The APA style for reference citations is similar to the second style of CBE citations. With the APA method, you include the author's last name, and the year of publication within parentheses, in your document; however, this method is slightly different. APA style for one author looks like this: (Duits, 1993). Notice the inclusion of the comma in the APA style and the absence of it in the CBE style. For two authors, the citation is (Duits & Smith, 1994)—again notice the comma and the inclusion of an ampersand (&) instead of the word *and*. For a citation with multiple authors, the Latin term *et al.* replaces the words *and others*—(Duits et al., 1995). Multiple sources should be included within parentheses in alphabetical order, separated by semicolons—(Duits, 1993; Hoffman, 1999). As with the second style of CBE citations, the reference list for APA citations should be in alphabetical order at the end of the document.

The MLA style for citations includes the author's last name and the page number on which the referenced information can be found. For example, you could have a reference of (Duits 134) within the body of your document. In this case, a reader would look at your reference list, find the article by Duits and know that the cited information is on page 134 of that document. With two or three authors, the word *and* is used within the citation: (Duits and Smith 134–37). Use et al. for more than three authors. If you use the author's name in the body of your text, just include the page number for the reference within parentheses: "Duits (134) found that. . . ." Multiple sources are separated by semicolons within the parentheses: (Duits 134; Hoffman 165). When you use the MLA citation method, the references appear at the end of the document in an alphabetical list. If more than one reference is cited for the same author, you include an abbreviated title of the work in quotations within the citations. For example, you might have (Duits "Education" 154) and (Duits "Combined Research" 121) within the same document.

In addition, some engineering professional societies, such as the IEEE, have established their own style manuals for citing references (www.computer.org/author/style/index.htm). For example, the IEEE specifies that the references should be cited in the document in the order in which they appear, and the reference number should be enclosed in brackets (i.e., [2]). The references cited are then included in the list of references found at the end of the document in the order in which they first appeared in the document. Thus, when writing technical documents for specific audiences, you must determine their preferred style for citing references and be sure to comply to their standards.

Sometimes you will be required to use footnotes to cite references within your document; however, this writing style is generally not preferred for technical documents. In most word processors, you can automatically create footnotes at the bottom of the current page or as endnotes at the end of the document. The presence of a footnote is generally shown by a superscript within the body of the text. For instance, "Duits[1] found that . . ." might occur within the text. At the bottom of the page would be the full reference (see the following Mechanics section for a description of a full reference) with a superscript—[1]Duits, E. J. (1999) . . .—or the number followed by the full reference—1. Duits, E. J. (1999). . . . Generally, the superscripts begin with *1* and increase sequentially throughout the document. If you repeat a reference on a later page, it will have a different number but you may use a shortened version of the reference (e.g., author(s), date, and shortened title).

Mechanics. When you are listing cited references in the section at the end of your document or in a footnote, the format will vary slightly depending on the citation style you chose and on whether the work is a book or an article. Each type of reference includes the last name and initials (or first name) of all the authors, the title of the work, the publisher, and the publication date. For articles, include the journal name and its volume and number. Online or electronic sources require additional information. In general, a reference should include enough information so that someone reading your document can go to the library or go online to find the work in question if desired. Some general rules for and comparisons among APA, CBE, and MLA reference list entries are as follows:

- For APA and CBE style, use an author's last name and first initial; for MLA style, use the author's entire first name.

- APA and MLA reference entries have a hanging indent on the second line; CBE reference entries are numbered, and succeeding lines of an entry align, or hang, on the first word after the number.
- In MLA and APA styles, reference entries are listed alphabetically by the author's last name; in CBE style, reference entries are generally listed in the order in which they appear in the document.

Some sample formats for citing references are listed next according to the type of the reference work and the citation style. For more complete descriptions of reference and citation styles, refer to the appropriate APA, CBE, or MLA handbook. If you are unsure about what style a publisher or your instructor wants you to use, ask: APA, CBE, and MLA formats are the major styles, but others exist, and publishers (or instructors) may want you to use their own preferred style.

Book References

APA   Earle, J. H. (1987). *Engineering design graphics*. Reading, MA: Addison-Wesley.

CBE   Earle JH. 1987. Engineering design graphics. Reading (MA): Addison-Wesley. 265 p.

MLA   Earle, James H. *Engineering Design Graphics*. Reading, MA: Addison-Wesley, 1987.

Journal Articles

APA   Nusholtz, G. S., Wu, J., & Kaiker, P. (1991). Passenger air-bag study using geometric analysis of rigid-body motion. *Experimental Mechanics, 31*(3), 264–270.

CBE   Nusholtz GS, Wu J, Kaiker P. 1991 Sept. Passenger air-bag study using geometric analysis of rigid-body motion. Exp Mech 31(3):264–70.

MLA   Nusholtz, George, James Wu, and Paula Kaiker. "Passenger Air-Bag Study Using Geometric Analysis of Rigid-Body Motion." *Experimental Mechanics* 31.3 (1991): 264–70.

Organization as Author

APA   American Institute of Steel Construction. (2001). *Manual of steel construction: Load and resistance factor design* (3rd ed.). Chicago: Author.

You may need to reference a large number of online or electronic resources, and when you cite such sources, some of the information in the reference entries will differ from that for a book or journal article. Examples of the type of online references you may need to provide include a government Web site, an online journal, a revised Web site, an article from conference proceedings on a CD-ROM, or information obtained from a CD-ROM. We cannot cover all the possibilities in this book, but the general rule still stands: *you should include enough information so that someone reading your document can go to the electronic resource and find the work in question.* If you are following APA, CBE, MLA, or another style, you should consult the appropriate style manual for styling reference entries for electronic resources. Another source that contains further information on this topic, listed in the References section of this chapter, is Barber (2000). Some examples of citations you may need to make in this electronic age are as follows:

A Web Site (Barber 2000)

Bodleian Library. Map Room. (Rev. 12 Mar. 1999).
http://www.bodley.ox.ac.uk/boris/guides/maps/ (22 Mar. 2001).

In the preceding citation, the first parenthetical date is the most recent revision of the Web site. If no revisions or modifications to the Web site are listed at the site, omit this element. The second parenthical date is the date on which you accessed the site.

A Government Web Site (Barber 2000)

Library of Congress. (1998). *American Life Histories: Manuscripts from the Federal Writers' Project, 1936–1940.* (Rev. 19 Oct. 1998).
http://lcweb2.loc.gov/ammem/wpaintro/wpahome.html (25 Apr. 2000).

An Online Journal (Barber 2000)

Wysocki, A. F. (1998). Monitoring order: Visual desire, the organization of Web pages, and teaching the rules of design. *Kairos,* 3(2).
http://english.ttu.edu/kairos/3.2/features/wysocki/bridge.html (19 Mar. 1999).

Note the similarity between the citation method for an online journal and that for a print journal, except the page numbers are replaced by the Web address for an online journal.

Conference Proceedings on CD-ROM

Bulleit, W. M., and Rosowsky, D. V. (1998). "Reliability Analysis of Light-Frame Walls Subjected to Wind Loads," *Structural Engineering World Wide,* Elsevier, Paper No. T206-2 (CD-ROM).

In this reference, the parenthetical *CD-ROM* is optional but will clarify to the reader that the reference is from a CD-ROM.

Formats other than those illustrated in this chapter may be used. You should determine the specific format required by your instructor, your employer, your publisher, or any other appropriate source and style your citations accordingly.

3.4   Avoiding Plagiarism

According to *Webster's New World Dictionary,* the verb *plagiarize* is defined as "to take (ideas, writings, etc.) from (another) and pass them off as one's own."

Interestingly, the word *plagiarism* stems from the Latin *plagiarius,* meaning "kidnapper." Plagiarism is both illegal *and* unethical (some things are one but not the other), and you should ensure that your documents are plagiarism free. If you steal another person's wallet at gunpoint, the crime is clearly understood. Plagiarism, however, is not always so readily recognized since it can be either unintentional or deliberate. We have seen cases in which students inadvertently plagiarized (by not rewording passages sufficiently), and we have seen deliberate acts of plagiarism. Both cases are highly unethical and should be avoided at all costs. Plagiarism reflects badly on you as a professional and could result in legal or disciplinary action if it is too blatant or damaging.

How do you avoid plagiarism and still cite others' work? Hotchkiss and Nellis (1988) pointed out that citing others' work is essential but does not give you carte blanche to copy their words. You must paraphrase to present others' work ethically. These researchers presented the following general rules for correctly paraphrasing (and thus not plagiarizing) others' work:

■ Do not change technical terms unless you fully understand them and can devise an equivalent substitute. For example, you can usually substitute *proportional limit* for *elastic limit* in a technical document, provided you know what they both mean. If, however, you do not know what is meant by *elastic limit,* you should retain the original wording.

■ Change the sentence structure and the choice of wording for nontechnical terms within the text being paraphrased. You can also move elements from one sentence to another.

■ Modify the sequencing of the information both within and between sentences.

■ If you do use an exact phrase from an author you have cited, be sure to include it within quotation marks ("exact words of author") so that readers know these words are not yours but the cited author's.

The following examples of plagiarism and paraphrasing were developed by Michigan Tech Associate Professor of Biological Sciences Dr. Ronald K. Gratz and used with his permission:

Original (from Gratz 1984): "Bilateral vagotomy resulted in an increase in tidal volume but a depression in respiratory frequency such that total ventilation did not change."

Unacceptable Plagiarism: "Gratz (1984) showed that bilateral vagotomy resulted in an increase in tidal volume but a depression in respiratory frequency such that total ventilation did not change."

This sentence is identical to the original except the author is cited.

Unacceptable Plagiarism: "Gratz (1984) showed that bilateral vagotomy produced an increase in tidal volume and a depression in respiratory frequency so that total ventilation did not change."

Changing a few words does not alter the fact that this sentence is still substantially the same as the original.

Acceptable Paraphrasing: "Gratz (1984) showed that following bilateral vagotomy the snakes' tidal volume increased but their respiratory frequency was lowered. As a result, their total ventilation was unchanged."

Although the same information is presented, the sentence structure and word order have been substantially altered.

Unacceptable Paraphrasing: "Gratz (1984) showed that following vagotomy the snakes' lung volume increased but their respiratory rate was lowered. As a result, their breathing was unchanged."

Dropping the adjective bilateral *alters the sense of the experimental technique.* Lung volume *is not the same as* tidal volume, *and* breathing *is not the same as* total ventilation.

3.5 Exercises

1. Find a Web advertisement or article that you believe portrays misleading information. Write a memo to your instructor describing what you believe are the ethical violations. Turn in a hard copy of the page along with your memo.

2. Visit the National Society of Professional Engineers Web site (www.nspe.org) and read the engineers' code of ethics. In a memo to your instructor, paraphrase the code. Include as part of your memo a citation for the Web page from which you gleaned your information.

3. In a small group, visit the engineering ethics Web page at Texas A&M University (www.ethics.tamu.edu). Examine one of the case studies, and discuss it as a group. Collaboratively write a memo to your instructor about the case, ensuring that both sides are presented. Include the conclusions drawn by your group about the case. Consider using bottom line up front (BLUF) when writing your memo.

4. In a small group, find your university's code of conduct for computer use. Discuss the code and determine whether specific items in it were surprising to you. As a group, compose a memo to your instructor outlining the most important points of the code and describing your group's reaction to it. Have your group leader e-mail the finished memo to your instructor as an attachment.

5. Select one of your textbooks and write citations for it using CBE, APA, and MLA styles.

6. Find a journal article in your library and write citations for it using CBE, APA, and MLA styles.

3.6 **References**

American Psychological Association. 2001. *Publication manual of the American Psychological Association.* 5th ed. Washington, DC: American Psychological Association.

Barber, M. M. 2000. *The Longman guide to Columbia online style.* New York: Longman.

Gibaldi, J. 2003. *MLA handbook for writers of research papers.* 6th ed. New York: Modern Language Association.

Gratz, R. 1984. Effect of bilateral vagotomy on the ventilatory reponses of the water snake, *Nerodia sipedon. American Journal of Physiology* 246:R221–7.

Hotchkiss, S. K., and M. K. Nellis. 1988. Writing across the curriculum: Team-teaching the review article in biology. *Journal of College Science Teaching* 18: 45–47.

Style Manual Committee, Council of Biology Editors. 1994. *Scientific style and format: The CBE manual for authors, editors, and publishers.* 6th ed. Chicago: Cambridge Univ. Press.

Walpole, R. E., R. H. Myers, S. L. Myers, and K. Ye. 2002. *Probability and statistics for engineers and scientists.* 7th ed. Upper Saddle River, NJ: Prentice Hall.

4

Personal Communication

Engineers write some form of communication virtually every day. Therefore, as an engineering student, you must understand the various forms of written communication you might be expected to produce, critique, or evaluate. Writing, especially technical writing, is a skill best developed through continuous practice. Consequently, you should strive to improve your writing skills and to practice them regularly. At first, writing may seem to be a painstaking experience, with the blank piece of paper staring you in the face a long time before you begin to put your thoughts down on it. However, with practice and experience you will be able to write increasingly well. In this chapter, we describe many forms of personal written communication you might use as an engineering student or as a professional. In subsequent chapters, we discuss and illustrate other forms of technical written communication.

As an engineering student or as a practicing engineer, you will often communicate with others directly through letters, memos, and e-mail. With direct forms of communication, the recipient and the sender are clearly identified by name on the document. In the following sections, we describe the most common forms of direct communication you will encounter as an engineer.

4.1 Letters

Function. In the business world, letters are used for direct communication with one other person. Unlike personal letters, business letters are typically brief and to the point. Frequently, such letters are single-purpose documents written to create a paper trail throughout the engineering design process. For example, if you are an engineer in a consulting firm and are responsible for the design of a city parking deck, you would probably start by asking your client to express, in a written letter, the required capacity for the parking deck, its location, and other characteristics. In this way, everyone has a clear idea of the project requirements, and, if questions arise later, a written record of the stated initial requirements exists. In other cases, letters inform clients, civil servants, salespeople, or others with whom you may be working of progress to date or decisions made.

In some instances, you may need to send a copy of a letter to a person other than the primary recipient. In this case, you should indicate on the letter that another person is receiving a copy of it so that the primary recipient is aware that this person was also informed. In other instances, you may include additional documents, called *enclosures,* with your letter. If your letter includes an enclosure, make sure you tell the recipient about the enclosure attached to the letter so that he can contact you if it was inadvertently left out of the envelope.

Form. In the business world, letters are usually written on company stationery (letterhead) that includes the company name, the company address, and possibly the company telephone and fax numbers. In some cases, e-mail and Web addresses may be part of the information provided on the company letterhead. When using a word processor to write letters, you can either print or photocopy the text of the letter onto the company letterhead. Figure 4.1 shows a typical business letter.

Mechanics. Because the letterhead already contains the name and address of your company, you need not include this information in the letter, as you may have been taught to do when you are creating a personal letter. However, for documentation purposes, you should always include the date on your letters. The date on the letter can be inserted at the left margin of the page, as shown in Figure 4.1. Alternatively, it can be inserted about 2 to 3 inches from the right margin.

Include the name, title, and address of the letter recipient below the date and left justified. If you don't know the person's name, you can address the recipient as "Dear Sir/Madam" or "To Whom It May Concern"; however, many businesspeople dislike the use of these phrases, considering them either archaic or impersonal. If you do not know the exact name of the person to whom you are sending the letter, you should use a representative title for the group to which the reader belongs. For example, you could use "Dear Project Manager" or "Dear Customer."

The information in the body of the letter should be brief and to the point. When you are writing to someone you have never met or spoken with, customary practice is to introduce yourself at the beginning of the letter. For example, you may begin the letter by saying something such as "I am a senior project engineer with the Braun Corporation in charge of the construction of the new city sewage facility." If the letter is a written response to an earlier oral communication, you might start with something such as "As you may recall from our phone conversation of January 15. . . ."

You should always end the letter with a closing such as "Sincerely," followed by three blank lines for your signature, and your name and title typed underneath

---

XYZ Consulting   1055 Washington Boulevard   Fort Wayne, IN 51001

January 2, 2005

Ms. Helen Johnston
Staff Engineer
United Pump Technologies
120 Avenue G
Evanston, Illinois 53321

Dear Ms. Johnston:

As you requested by telephone last week, I am sending you some information, enclosed with this letter, about the reliability of United Pump Technologies Model SR151. Our company has used this particular pump model in numerous well and water system designs in the Fort Wayne area. We have had success with the pumps in most installations. However, in certain locales they have been known to "freeze up" and had to be replaced. Most of the failed installations have been in sandy soil types, although this finding is not true of all cases. I am also shipping you one of the pumps that failed.

Please let me know if I can be of further assistance.

Sincerely,

Bob Griffin

Bob Griffin
Project Engineer

Encl: Model SR151 Field Report

Figure 4.1. Sample business letter

for legibility. The letter closing should be aligned with the date. Thus, it should be at the left margin of the page or placed 2 to 3 inches from the right margin.

If someone else has also received a copy of the letter, include this information near the bottom of the page, left justified. The correct format for including such information on your letter is as follows:

CC: D. Johnson

The abbreviation *CC:* stands for "carbon copy" and reflects an earlier era when photocopiers were not available and carbon copies of documents were the norm.

Similarly, if an enclosure accompanies your letter, signify that this is the case by including *Encl:* and a short description of the enclosure at the bottom of the letter, left justified:

Encl: Program brochure

If you are writing a business letter or cover letter and aren't using letterhead, make sure your contact information is included on it. One way you can include your contact information is at the top of the letter on the right side of the page (upper right corner of the page), above the contact information for the person who will be receiving your letter. Another way you could include your contact information is directly below your signature line. For example, for the letter shown in Figure 4.1, the following could consitute the letter closing if no company letterhead were available:

Sincerely,

Bob Griffin

Bob Griffin
Project Engineer
XYZ Consulting
1055 Washington Boulevard
Fort Wayne, IN   51001

4.2   Memos

Function. Much internal written communication in an organization involves using *memoranda,* usually referred to as *memos.* Memos are not so formal as letters, but they are equally important in the documentation of an engineering design project. A company memo form (similar to the formal company stationery or letterhead) is not required, but it may be used depending on company policy. Most memos are single purpose, brief, and to the point. Copies of memos are commonly given to one or more people in your company so that everyone working on a given project receives relevant information as needed. Some specific functions of memos are as follows:

- To act as internal organization letters
- To document work accomplished
- To document oral discussions
- To disseminate information to large numbers of people
- To record technical information

Form. Memos are seldom longer than several pages and are often less than one page long. When composing a memo for use in your company or for use in one of your university classes, you should try to answer three basic questions:

1. What was the question?
2. What is my answer?
3. Why should you believe me?

Figure 4.2 shows a typical memo format. Since memos are written to individuals within your company or office, you do not need to include the recipient's address, and the writing style is not usually as formal as that in a letter. Personal

```
┌─────────────────────────────────────────────────────────────────────────────┐
│                                  MEMO                                         │
│                                                                               │
│   TO:        Maria Sanchez, Project Engineer                                  │
│   FROM:      George Chou, Project Manager      GC                             │
│   DATE:      January 1, 2005                                                  │
│   RE:        Field Inspections                                                │
│   CC:        C. Hoffman, J. Burke                                             │
│                                                                               │
│   As you requested, I have assigned Claire Hoffman and John Burke to do the   │
│   field inspections for the upcoming construction of the Brighton County      │
│   water treatment facility. Both these individuals have extensive field       │
│   experience and they should perform well on this project.                    │
│                                                                               │
│   The Brighton County Public Works Department expects us to be diligent in    │
│   our inspections. They are concerned that the contractor may not be fully    │
│   up to date on some of the new treatment equipment that we specified. Be     │
│   sure to warn Claire and John to take special care during the installation   │
│   of this equipment. We should be okay since we have two sets of eyes keeping │
│   track of the construction.                                                  │
└─────────────────────────────────────────────────────────────────────────────┘
```

Figure 4.2. Sample memo format

pronouns such as *I* and *you* are generally acceptable. Notice that in the memo shown in Figure 4.2, the three basic questions were successfully answered:

1. The basic "question" was that George Chou was asked to assign field inspectors for the project at hand.
2. The answer to the question was that Claire Hoffman and John Burke were assigned this task.
3. The fact that these individuals have extensive field experience shows why they were a good choice for the task.

Mechanics. Typically, five standard lines on a memo are included in its heading:

1. TO:
2. FROM:
3. DATE:
4. RE:
5. CC:

In addition to these lines, some companies include a *REF:* line in the heading area to give reference to the specific project or previous memo that the current one was written to address. The order of items 3 and 4 is sometimes reversed; however, items 1, 2, and 5 are almost always included in these positions on the memo heading. In other cases, item 3 (the date) is put at the top of the page to facilitate filing for project documentation (i.e., all memos related to a specific project are included in chronological order in the file). When you are inserting the required information on these heading lines, you should use the *Tab* key before inserting the information so that all fields in the heading align.

Include the name and the title (optional) of the recipient of the memo on the *TO:* line of the heading. Sometimes you will need to include more than one name and title, if your memo will be sent to more than one individual. Insert your name and title (or more than one name if you are writing the memo with someone else) on the line labeled *FROM:*. In memos, as in letters, you should include the date (on the *DATE:* line) for documentation purposes. After *RE:* (short for "in regard to"), insert a brief description of the purpose of the memo. The line labeled *RE:* is sometimes labeled *SUBJECT:*. Either of these two styles is acceptable, but you should use the style preferred in your company.

Often you will need to copy a memo to individuals other than the direct recipient or recipients of the memo. For example, if you are writing a memo to a colleague but want to make sure your boss is kept in the loop, you would probably send the memo *TO:* your colleague and *CC:* your boss. When copying a memo to other individuals, include the names of everyone who should receive a copy of the memo on the line labeled *CC:*.

The body of the memo follows the heading information. The paragraphs of the memo are usually block style (i.e., no indentation), although indented paragraphs are also acceptable. Since the memo is a short document, it is easier to read if paragraphs are kept short (i.e., less than about 60 words). A common plan for the body of a memo is as follows:

- Tell the reader the subject and purpose.
- Develop the subject.
- Discuss conclusions, decisions, and recommendations.

In some companies, the preferred style is for the conclusions, decisions, and recommendations to be near the beginning of the memo. In this case, include a sentence that states the conclusions in the first sentence of the memo, then include the other items.

Signing memos with a complete signature, as you do with letters, is not usually important, although a good practice is to initial the memo near your name.

4.3   Cover Letters and Memos

Function.   Cover letters and memos are a special type of person-to-person communication and are usually included when you are attaching a document to the letter or memo. A cover letter will point out the highlights of the attached document and will describe its contents. Cover letters are somewhat like previews of coming attractions, in which you inform the person receiving the letter about what to expect in the attached document. Documents distributed inside a company might have a cover memo. This is simply a cover letter in memo form.

Probably one of the most common uses of a cover letter for college students is as an accompaniment to a résumé during a job search. Since a résumé cover letter will be your first introduction to a potential employer, it must be well written, error free, and interesting. With this type of cover letter, you are trying to convince the recipient to study your résumé and invite you for an interview. Cover letters that accompany résumés are covered in more detail in Chapter 10.

Form. The form for a cover letter is similar to the form illustrated in the previous section of this chapter. In general, cover letters should rarely be more than one page long. Recall that the purpose of the cover letter is to convince the reader to read the enclosed document. If the cover letter is too long, the reader may be discouraged from reading further.

If your cover letter is accompanying your résumé, it will typically not be written on company letterhead, so you should make sure your contact information (i.e., address, phone number, e-mail address) is included either on your résumé or as part of the cover letter.

Mechanics. Cover letters are typically addressed to a specific person, rather than "To Whom It May Concern" or "Dear Sir/Madam." If you don't know the name of the person who should receive the materials, do some research to determine the name of the proper person within the organization. Cover memos follow the same form as regular business memos and are reserved for individuals within your company.

For a résumé cover letter, use the opportunity to point out some unique features of your qualifications so that the reader is interested in looking at your résumé to learn more about you.

Example

An example of a type of cover letter often used by university students is shown in Figure 4.3. This letter is the type used to convey the results of a student design project to a corporate sponsor.

4.4   Electronic Mail

Electronic mail, e-mail, is a pervasive part of university life as well as of corporate America. E-mail often takes the place of letters, memos, phone calls, and sometimes even meetings. In the 1980s, e-mail was a novelty. It was fun and engaging. Just a few individuals had access to e-mail. Today, e-mail is usually viewed as a chore that must be contended with in the work environment. In this section, we discuss some of the issues regarding e-mail that you must consider for effective communication. The specific tool you use to compose, send, and receive e-mail will depend on the software available to you. We recognize that many of you may already be "experts" at sending and receiving e-mail; however, most of your experience with e-mail has likely been sending e-mails to and receiving e-mails from your friends and family—when formality and clear communication has not been an issue. The following discussion about e-mail is included to help you think about some issues you may not have considered in your past e-mail experience.

Function. Similar to letters and memos, e-mail is used for direct communication between two individuals or among groups of individuals. E-mail is quicker than sending a letter or a memo, is cheaper than using the U.S. Postal Service, and is

May 1, 2005

Mr. Paul Anderson
Senior Project Engineer
Paper Products Division
Helvetica International
Appleton, WI 54400

Dear Mr. Anderson:

Enclosed please find a copy of our design report titled "Evaluation of a Disposable Blood Pressure Cuff for Commercialization." This report highlights our findings about marketing, manufacturability, and economic considerations for a paper-based disposable blood pressure cuff. Also included in the report are three possible preliminary designs for such a device. Each design has advantages and disadvantages as outlined in the report; however, after careful consideration of all factors, our design team makes the following two recommendations:

1. We do not believe that this product will be profitable in the long run and recommend that it not be pursued further.
2. If your company chooses to proceed with this product development, we recommend that preliminary design B be further explored (it is the most versatile and cost effective, as described in our report).

Our team members appreciate all the hard work you have done for us during the past year. Thank you for everything you have done for us and this project. If you have any questions in the next few weeks, feel free to contact us.

Sincerely,

Emily Rossi

Emily Rossi
Senior Project Team Leader
Chemical Engineering Department
Michigan Technological University

Figure 4.3. Sample cover letter

better for sending out mass mailings. For these reasons, e-mail has rapidly become the preferred mode of communication for most professionals. E-mail has somewhat replaced the need for letters and memos; however, letters and memos will probably still be used for documentation purposes into the forseeable future.

Form. The form for an e-mail message is a hybrid between that for a memo and that for a letter. As in a memo, heading fields should be filled in for an e-mail message. When composing e-mail, you include the recipient's e-mail address on the *To:* line, the subject of the e-mail on the *Re:*, or *Subject:*, line, and, on the *Cc:* line, the e-mail addresses of other recipients who should be aware of the message content. The date and the sender's name will automatically be supplied by the tool when the message is sent. Note that you can list several e-mail addresses, separated by commas or spaces, on the *To:* and *Cc:* lines to send your message to a large number of people simultaneously.

Unlike memos, however, e-mail messages should almost always start with a salutation and end with your "signature," just as you start and end letters. For the e-mail salutation, you sometimes use *Dear* in front of the recipient's name, especially if you are writing to someone you do not know personally. Alternatively, you can just use the person's name as the salutation for the e-mail message, especially if you are writing to someone with whom you are well acquainted. When "signing" your e-mail, you include just your first name for a message to someone you know well and both your first and your last names, as well as your official title on the line below your name, for a message to someone you don't know. Some e-mail tools allow you to include a signature footer that is automatically placed at the end of any e-mail message you send. The footer usually includes your title, your contact information, and any personal message you want to convey.

Mechanics. When you are sending e-mail to a friend or a colleague with whom you have a relatively casual relationship, generally acceptable practice is to use all lowercase letters and not to follow all grammatical rules strictly. You might use certain abbreviations in an e-mail to a friend (e.g., *btw* means "by the way," *imho* means "in my humble opinion," *fyi* means "for your information," *$.02* means "my two cents" when you are stating an opinion).

When you are writing a formal e-mail to your boss or someone you don't know well, you should use the same care as when you are preparing a formal letter or memo. Rules of grammar and punctuation should be followed; words should be capitalized as necessary; your writing style should be formal. If you are writing an e-mail to one of your university instructors, include your last name on your signature line. A faculty member might have 10 students with the name *Jessica* among all the students in her classes. You want to make sure she knows which student sent her the e-mail since she may not be able to determine who you are from your e-mail address. (*Hint:* If you are writing your instructor about missing a class, be sure to tell her which class you will be missing and why. Also, never ask your instructor if "anything important" will be occurring in her class on a day you plan to miss; she probably thinks something important occurs there every day.)

Certain conventions have evolved about proper e-mail communication. If you want to emphasize a word or a phrase, surrounding it by underscores (_) or asterisks (*) is the equivalent of underlining or boldfacing it. Thus, typing *_opinion_* or **opinion** would place special emphasis on the word *opinion* in an e-mail message. Writing something in all capital letters is the equivalent of shouting at someone, so typing *OPINION* in an e-mail would stress your point even more strongly. Generally, you should avoid using all caps in professional e-mails unless you believe doing so is necessary. Many e-mail tools now enable you to include these character treatments as part of the e-mail message; however, not everyone uses an e-mail tool that can interpret them correctly. When in doubt, err on the conservative side and send plain text e-mail messages.

Some people choose to include *emoticons* in their e-mail messages to convey emotions that might not be understood through text alone. For example, if you write a message that offers a person constructive criticism, you might want to include a smiley face to make sure your reader does not take offense at your message. If you were talking to someone face to face, your body language could convey that you are offering the criticism as a friend and not to be cruel. Since an e-mail message cannot show your body language, if you include emoticons, you might be able to diffuse a potentially awkward situation. Some examples of the more commonly used emoticons are these:

:-) Smiley face

:-(Frowning face

;-) Winking face

The *Cc:* function of the e-mail tool is used to send copies of the e-mail to people other than the primary recipient. Another function is the use of *Bcc:* (which stands for "blind carbon copy"). You send a *blind carbon copy* to a person when you want that person to receive a copy of the e-mail but you don't want the primary recipients to know that this person is receiving a copy of it. Using this function may seem kind of sneaky, and some people may find it unethical, but sometimes you may find using it necessary. For example, suppose you are working on a team design project in one of your classes and a teammate is not participating in the team's work. You might send an e-mail to the teammate asking if a problem is preventing him from meeting with the team. In this case, you might want to send a blind carbon copy to your instructor so that she is aware of the situation. If you send someone such as your instructor a blind copy, you should follow up with an e-mail to that individual telling her that she received a blind copy and just letting her know that you did so to keep her in the loop, not that you expect her to do anything at that time. This way, the recipient of the blind copy will not say the wrong thing to others who may have received the e-mail or a *Cc:* of it.

E-mail addresses consist of three basic parts: a *user ID,* followed by the *"at"* symbol (@), followed by the *server location* of the user ID. For example, *cduits @ csu.edu* means the person's user ID is *cduits,* and *csu.edu* is the name of the server that handles e-mail for cduits. Many times within a company or a university campus, where all e-mail is handled by a central server, you do not need to include the *@ server* as part of the recipient's e-mail address (i.e., if you are a student whose home server is csu.edu, you could just use cduits as the address to which you are sending an e-mail message). Some companies may have a firewall for network security, which might mean you must use special considerations when you are sending e-mail to someone on the other side of the firewall. A good idea is to understand the e-mail conventions in place in your situation to ensure that you do not breach security or policy.

If a person with whom you regularly exchange e-mail has an e-mail address that is cumbersome to use or difficult to remember, you should probably create an alias or a nickname for the person. Most e-mail tools have the ability to create aliases or nicknames easily. An *alias* or *nickname* is usually a shortened version of a person's e-mail address that can be typed quickly or can be easily remembered. For example, if you regularly exchange e-mail with a person whose e-mail address is trotter44 @ usma.edu, you might want to create an alias of *steve* for this person. Then, when you want to send an e-mail to him, you need only type *steve* on the *To:* line of the e-mail, and your tool will automatically readdress the message to his official address.

You can also use the alias or nickname function of your e-mail tool to create substitutes that represent several e-mail addresses at once. This feature is especially useful for work on group projects. For example, if you are working on a semester project and five individuals are on the design team, you would not want to include five addresses on the *To:* line each time you wanted to send out a meeting reminder. Your alias or nickname for the team could be something like *project_team*. If you then include *project_team* on the *To:* line of an e-mail, your e-mail tool will automatically direct your message to each of the five individuals who make up the alias or nickname.

Another function of e-mail is the easy exchange of documents or other types of electronic files. Electronic files can be attached to an e-mail message and sent along with your message to the recipient. Thus, two or three people can work on writing a paper together with relative ease. One person could work on his portion of a paper and send it on to a coauthor so that she can pick up where he left off and complete the paper. Digital photographs showing specimen failures, for example, could be sent to a colleague who might have some insight into the cause of the failure.

As with any productive tool, abuses can occur. Computer viruses are often sent as e-mail attachments. If you open an attachment that contains a virus, you could destroy the contents of the hard drive, which would require you either to reformat it or possibly even to buy a new one. Opening an attachment on your university student account can have even more dire results. Since your university computer system is likely networked, one virus could disable several servers, destroying the data of thousands of users. For this reason, you should *never* open an e-mail attachment unless you are sure about its contents. You should especially never open an attachment sent to you by someone unknown to you.

One other type of e-mail abuse that has erupted is *spam*. Some people's definition of *spam* might vary, but typically it is thought of as any type of mass, unsolicited e-mail you receive. Many companies send spam to customers, especially if the customers have purchased anything online from them previously. Spam can also originate with other organizations or individuals who think this type of mass mailing will help them get their message across. Spam is annoying and clogs e-mail systems similar to the way junk mail clogs the postal system. You should avoid sending out mass e-mails that others may perceive as spam.

Examples

Figures 4.4 and 4.5 show examples of formal and informal e-mail messages.

4.5 Electronic Lists

Function. An electronic list is usually used for communication among large or small groups that need to have a forum for discussion of ideas or for regular communication. A list is useful for sending out e-mail to large groups simultaneously.

Dear Mr. Sullivan,

Thank you for offering to assist our senior design project class and for agreeing to make a presentation to us about your Cuff-Guard product next month. We all look forward to learning about your product and the different aspects you dealt with in producing a new product for market. Before we begin what we hope will be a fruitful partnership, our class would like to know if you would like to be an official mentor to our class for the coming year? Our class consists of teams that are mostly self-explanatory. They include the production team, testing/ prototype team, ESH (environmental, safety, and health) team, marketing team, Basis for Interest documentation team, and public relations team. Each team has its own leader and makes a list of personal and team objectives for each semester. If you'd like to be a mentor, which team would you prefer to be a part of or a mentor for?

To assist you in tailoring your upcoming presentation to our class, we offer the following set of questions as "food for thought." Our testing team wants to know how you tested your Cuff-Guard. Did you use a model arm when taking a blood pressure or did you not worry about that? If so, how did you reproduce a consistent blood pressure (human, model)? How did you calibrate your testing techniques to meet the FDA regulations? What were your standards for marketing? The production team is interested in the answers to the following questions: What materials did you choose, how did you choose them, and how did you assemble the different layers? Also, what kind of equipment did you use to produce the Cuff-Guard? Finally, the Basis for Interest documentation team would like to know if you signed a BFI document for anyone for the Cuff-Guard?

As always, if you have any questions about our senior design team, being a mentor, or our project teams, feel free to e-mail our adviser, Dr. Tony Rogers (trogers@mtu.edu), or me. Thank you very much for your time, and we look forward to hearing from you and to meeting you in the coming month.

Sincerely,

Emily Rossi
Senior Project Team Leader
ejrossi@mtu.edu

Figure 4.4. Formal e-mail message from student design team to corporate partner

---

Katy,

It was great talking to you on the phone last week. I'm glad to hear you are doing well and seem to have recovered from our "strenuous" time at the recent ASM conference. This e-mail is just a friendly reminder that you promised to send me some information about testing you had completed on the austenite alloy we spoke about. If you can also include photos from your SEM, that would be great! See you at the ASM conference next year (I hope).

Lea

Figure 4.5. Informal E-mail message to colleague

If an e-mail message is posted to the list, all members on the list will receive that message. Your professor may have a list set up for each class he is teaching. In this way, he can send an e-mail describing changes in a homework assignment or a due date to everyone in the class simultaneously. Another use for lists is as a forum for discussion. Individuals can post messages to the list; others who are part of the list can then post their opinions on the topic to the list in response to the original message. The discussion can continue on a given topic until it's time to move on to the next topic or until no one has anything they want to add to the "conversation."

Form. Posting a message to a list is similar to sending an e-mail to an individual. You merely insert the "e-mail address" of the list on the *To:* line in your e-mail tool. You might include a salutation at the beginning of the posting, but not always. You should always include your signature at the bottom of your posting, along with your e-mail address. You include your e-mail address in case someone wants to discuss something with you *off-line* (i.e., a sidebar discussion that doesn't go through the list). If you are responding to something previously posted to the list, be sure to include enough of the original message so that others will understand what you are talking about. Customary practice is to include the portions of a previous posting in your message with an angle bracket (>) at the beginning of each line to indicate that it came from an earlier message:

>The soil borings on the proposed construction site indicate
>that there may be a drainage problem if the water main is
>put in the location shown on the drawings.

I think that the client would be best served if we were to reroute the water main to avoid this potential future problem.

In this example, the sentence beginning with "I think that" is in response to an earlier posting that started with "The soil borings." This convention of using angle brackets to indicate salient parts of previous messages is also widely used in responding to e-mail sent by individuals.

Mechanics. You will probably encounter several basic types of lists in your electronic communication. In a *closed-posting list,* only the members of the list can post to the list. With a closed-posting list, the server will check the user IDs of all incoming messages. If the sending user ID is not recognized as part of the list, the message will be returned to the original user and not posted to the list. Some lists are also *closed-subscribe lists,* which means you cannot join the list unless you are authorized to do so by the list owner. By contrast, anyone can post a message to or subscribe to an *open list.* With a *moderated list,* all e-mail messages must go through a single contact (i.e., only one or two people can submit messages to the list). If you want to post something to the list, you must first submit it to a moderator. She will then read it, decide whether it is appropriate for posting, and, if so, forward your message to the list. With an *unmoderated list,* anyone who has access can post messages to the list. Note that a list could be moderated and open, moderated and closed, unmoderated and open, or unmoderated and closed.

Another difference among lists is how replies are handled. With most lists, if you click the *Reply* button on your e-mail tool, your response will be sent to everyone who has subscribed to the list. Other lists are set up so that your reply goes only to the sender of the original message. Make sure you know how your list is set up. Many individuals have been embarrassed when they replied something in confidence to a friend only to learn that their "personal" message went to everyone on a given list. Another problem might occur in the way you set your AutoReply function for times when you are out of town. The AutoReply function is used to inform individuals who send you e-mail that you are unavailable and will get back to them when you return. If you set this function to reply to *All*, you will create an infinite loop of e-mail messages. You first receive a message from a list, AutoReply sends a message back to the list saying you are unavailable. Since you are a member of the list, you will also receive this message. Because you have received the reply, AutoReply will again send a message to the list saying you are unavailable, and so on. This loop will continue forever if uninterrupted, but, in the meantime, the other list members will have received countless messages, which will clog their mailboxes and be a general irritant.

4.6 Exercises

1. Write a memo to your instructor detailing what you believe are the computer skills training needs of students in your major.
2. Locate and read a technical article on a topic of interest to you. Write a memo to your instructor summarizing the most salient points of the article. Turn in the article along with the memo.
3. Write a letter to your university president about a problem you have encountered on campus (but don't send it).
4. Write a letter to your dean of students describing what you like best about your campus (but don't send it).
5. Write an e-mail to your instructor describing your most recent group meeting. In the e-mail, list tasks that resulted from the meeting and who is responsible for carrying out each task.
6. Write a memo to your instructor about campus parking problems. Write a brief e-mail to your instructor and electronically attach the memo to it.

5

Document Design

As an engineering student or professional, you will often be called on to report the results of a study or an investigation you performed. Technical information is typically communicated as a report or as a paper. When developing your document, you should pay careful attention to its overall appearance. If your document is poorly designed, it may be difficult to read and comprehend. A well-designed document engages readers and enables them to quickly and easily find the information they need. Common elements found in most types of reports are a title page, a table of contents, a list of figures, and one or more appendixes. Before we begin our discussion of the creation of technical reports in subsequent chapters, let's examine these common document elements. In addition, this chapter includes information about document design in the electronic age.

5.1 Tool Selection

Tool selection is important in the initial stages of preparing to create documents for technical communication. Just as you would never choose a screwdriver to drive a nail into a wall, the tool you use to convey your technical information should be appropriate for the task at hand. In the early 1900s, the tools for technical communication were fairly limited—pencil, ink, paper, typewriter, telegraph, and, possibly, telephone. Currently, dozens of tools are available from which to choose. For example, if you want to communicate with your supervisor about a problem you encountered on the factory floor, you might choose among stopping by her office to discuss the matter with her, calling her on the telephone to give her a "heads up," sending her a quick e-mail about the problem, writing a brief memo using word-processing software, jotting a quick note on a sheet of paper and taping it to her door, and faxing her a note or drawing with the problem outlined. In some cases, you may choose to do more than one of these (e.g., a quick phone call followed by a memo or an e-mail).

The choice of tool you use to communicate a problem to your boss will likely depend on several factors: the severity of the problem, the immediacy of the problem, your personality, your boss's preferred management style, and the need for documentation in the future. For example, if the problem is immediate and could lead to a dangerous situation on the factory floor, you would likely need to call or find your boss directly to discuss the situation. In this case, after the problem is solved, you would also probably follow up with an e-mail or a memo to establish a paper trail and to document the nature of the problem and its solution.

In most cases, your tool selection will be dictated somewhat by the technical communication task at hand. For example, a professor may require you to turn in a document with professional-looking graphs and charts that has been prepared with a word-processing package. In this case, you will most likely use the software to which you have ready access either on your personal computer or in your school computer lab. In some instances, if you are submitting a paper to a journal or even to a conference, you will be told exactly which software package to use when preparing your document. In addition, you will likely be told the exact margin sizes, font types, font sizes, and so forth to use in your paper. If you are sending an e-mail to your boss, you may have more than one e-mail tool on your computer system to choose from. If you are working on a document with a colleague, you may have to work with a word-processing software package that is not your first choice, depending on his access to a particular tool.

In some instances, however, you will be able to choose, from several options, the appropriate tool for the job at hand. This fact is particularly true when you are creating graphics, creating visual aids for technical presentations, and creating documentation for your design calculations. Considerations of tool selection for each of these are presented in subsequent chapters of this text.

5.2 Fonts

Function. When designing your document, you must organize it in a manner so that readers can easily find the information they seek. The document organization should also engage the reader through variation in font types when appropriate. Headings should be set apart from passages of text so that readers can detect changes in topics and thus be able to skip entire sections or concentrate on specific issues as desired. In reality, you will probably be the only person to read your document from cover to cover, which means other individuals need to be able to navigate through it, gleaning the information they require, with relative ease.

Sans Serif Fonts	Serif Fonts
12-point Helvetica	12-point Times New Roman
10-point Arial bold	*8-point Bookman Old Style, italic*
14-point Avant Garde	12-point Courier, underlined
10-point Tahoma	*10-point Palatino bold italic*

Figure 5.1. Various fonts, sizes, and treatments

Form. Many standard fonts are available with modern word-processing tools. Fonts are generally classified as either *serif* or *sans serif* (*sans* is the French word meaning "without"). Serifs are the short "dashes" that appear at the ends of the character strokes in many fonts. The most commonly used font for technical communication is Times New Roman, which is a serif font. Research has shown that serif fonts are generally easier for people to read. You may want to occasionally use sans serif fonts to set a section of text off from the main body as appropriate.

Font sizes are defined in terms of *point sizes*. An inch comprises 72 points, so a 36-point letter would be one-half inch high. Most documents should be created with a 12- or possibly a 10-point font. If you are writing for a professor, remember that she might be at an age when small print is difficult to read, so make sure you use a font size of 12 points for documents you prepare for her. Figure 5.1 shows several types of fonts, treatments, and sizes available with most word processors.

Mechanics. Typically, you should use only one font type for the main body of your document: excessive font changes are confusing and annoying and in the end detract from your ability to communicate effectively. The font size can be used to distinguish between sections of your document instead.

With current printing and document reproduction capabilities, documents can now be created that contain text of several colors. Again, be conservative in your use of colors. Too many colors will detract from the readability of your document. However, the use of some colors may enhance its appearance and enable you to direct attention to critical information. For example, you could use red type in a table that shows a project budget to indicate which categories are "in the red." When using colored type, be sure to choose colors that are easy to see: avoid yellow and magenta since these colors will likely serve only as irritants.

5.3   White Space

Function. White space is a necessary element of all technical documents, and judicious use of white space will ensure document readability and clarity. Graphic designers and communication professionals typically spend a great deal of time considering white space when designing a document. The documents you prepare as an engineer will usually be less complex; however, white space is still an important consideration in ensuring clear technical communication.

Form. *White space* on a page consists of anything that does not have text on it—including line spacings, margins, and indentations. If you are writing a document for a specific audience or journal, many of the decisions about white space will probably have been made for you.

Paragraph justification is one of the white-space considerations you will have to think about as you design your document. Four main types of justification are possible: left, right, centered, and justified. *Left justification* means each line of text is aligned on the left side of the page, and the right side of the column is ragged. *Right justification* means lines of text are aligned on the right side of the page, and the left margin is ragged. *Centered justification* means each line of text is centered on the page. *Justified text* means both the right and the left margins are aligned.

Another white-space consideration you will typically encounter is paragraph indentation. Typically, three types of indentation may be used. First, indentation of the first line of a paragraph is often used to signify its beginning. Second, no indentation is often preferred for technical documentation; however, this style usually means a blank line must be left between paragraphs so readers know when a new paragraph begins and the previous one ends. Third, a hanging indentation means the first line is justified at the left margin and all subsequent lines are indented a few spaces from there.

Margins and line spacing are also important white-space elements for you to consider when you are creating your document. Line spacing is generally either single or double; however, it can be customized in virtually limitless ways with modern word-processing tools. Margins are typically about 1 inch on all sides but may be altered as needed. Figure 5.2 illustrates the most common white-space elements found in engineering technical documentation.

Mechanics. Single line spacing is often easier to read than double spacing; however, if your reader will be making editorial comments on your document, double spacing is preferred. Indentation decisions should usually correspond with your line-spacing decisions. If you choose single spacing, either paragraphs should be indented or a blank line should be left between paragraphs to distinguish one from another. Double-spaced text almost always has indented paragraphs to show the beginning of each one. In some cases, the first paragraph in a section of text is not indented and the remaining paragraphs are. This style is particularly common in textbooks and technical papers.

Left justification is generally preferred for most types of technical documents. Centered justification is suitable for titles and headings, but not full paragraphs of text. Right justification is used sparingly in technical documents. With

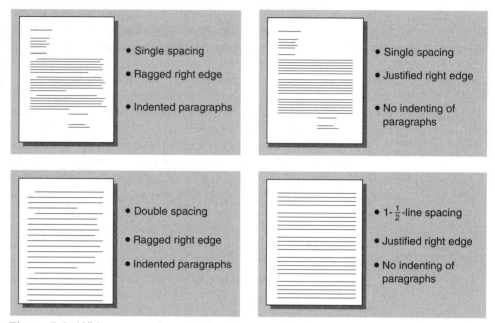

Figure 5.2. White space elements in document design

justified text, the character spacing is altered throughout each line to make all lines even. For this reason, justified text is generally more difficult to read and should be avoided unless there is a compelling reason to use it.

5.4   Headings and Subheadings

Function. Headings and subheadings are used in documents to distinguish sections of text from one another. They add to the clarity of the document by dividing it into manageable pieces. Headings also improve the readability of a document, enabling people to skip whole sections of text not critical to their understanding and to concentrate on sections where they need to do so. Imagine how difficult understanding this text would be if headings and subheadings weren't used to indicate topic changes and so forth.

Form. Headings should generally be of the same font family as the main body but may include special treatments or changes in size to set them apart from the remainder of the text. Documents typically have several levels of headings to distinguish among main sections or ideas within the body of the document. The level of a heading shows its relative relationship to others. First-level headings are the "top" level and are used to distinguish the main sections of text. Second-level headings are subordinate to a first-level heading and show how the main section is broken into smaller subtopics. Third-level and fourth-level headings are further subdivisions of second-level and third-level subheadings, respectively. Some types of treatments used for headings and subheadings include, but are not limited to, the following:

Boldfacing, Underlining, or *Italicizing* the headings to make them stand out

Increasing the Font Size

Numbering a first-level heading with a whole number (e.g., 4.0) and then numbering second-, third-, and fourth-level headings with decimals to indicate that they are subordinate to the main heading (e.g., 4.1 and 4.2 for second-level headings emanating from heading 4.0, and third-level headings of 4.1.1 and 4.1.2 for subtopics subordinate to second-level heading 4.1)

Centering or indenting the headings to separate them from the text

Mechanics. As a general rule, you should include no more than four levels of headings in a single document. Be consistent with your fonts and font treatments for headings at the same level. For example, if your first-level headings are 14-point Times, bold, and centered, make sure all your first-level headings are the same size (14 point), font (Times), and treatment (bold and centered).

Typically, font sizes decrease as the heading level increases. For example, if a first-level heading is 14 point, a second-level heading might be 12 point. As the heading level increases, it should also stand out less from the text. For example, third- or fourth-level headings are often treated as run-in headings (see Figure 5.3) rather than as freestanding headings.

Two main general rules should be followed when you are adding headings to sections of text within your document:

1. Never allow a heading to sit alone at the bottom of a page. Add spaces so that it begins at the top of the next page.
2. Don't follow a first-level heading immediately by a second-level heading without including a few sentences of text between the two headings. However, when third- or fourth-level headings are run-in headings, you may put them immediately after first- or second-level headings without text between the headings. For example, throughout this text, the third-level heading "Function" often follows immediately after a first- or second-level heading, without intervening text.

First- and second-level headings will become the basis for your table of contents for the document (discussed in Section 5.6). Third- and fourth-level headings typically do not appear in a table of contents. Figure 5.3 shows one possible division of a document by using headings and subheadings.

1.0 Section Heading

For the main sections of your document, you will likely want the section heading to be larger than the normal font size for the text. You should also consider treatments such as bolding or underlining to set the heading off from the text. In this example, the heading would be considered a first-level heading.

1.1 Subtopic Heading

In this example, subtopic heading 1.1 falls beneath the main section heading, indicated by a smaller font and a numbering scheme that denotes this organization. In this example, this subtopic heading would be a second-level heading.

1.2 Another Subtopic Heading

This section of the text is also subordinate to Section 1.0. Notice that the font size and treatment of this second-level heading matches that of the previous subtopic heading.

Third-Level Heading as Run-In. In this example, the next level of subheading uses a run-in scheme. Numbering this section as 1.2.1 might be considered tedious, thus a different scheme for setting this part off from the remainder of the text may be chosen.

2.0 Another Section Heading

This section is another main part of the document and requires a first-level heading. Note also that the use of margins helps to set the sections of the document off from one another.

2.1 Subtopic Heading

This second-level heading matches those used previously in the document. Note that this document contains only three levels of headings.

Figure 5.3. Headings and subheading scheme for a document

5.5   Title Page

Function. The title page provides an easy-to-read cover for a report: the pertinent information for the report is clearly visible at a glance. Thus, if the report somehow becomes buried in a pile of papers on a desk or is one of many your professor must read, it is easy to distinguish from others.

Form. Four main parts are essential ingredients of a title page. You may include one or two additional items; however, you should avoid clutter on a title page at all costs, so don't include too many extras. The four essential items are as follows:

1. Project title
2. Your client's name
3. Your name
4. Date of submission

The project title should be descriptive but as short as possible. Many novice writers make the mistake of using a title that is too long. In general, 10 to 15 words should be the maximum for a title. When composing your report title, try to avoid phrases such as "A Technical Report on . . ." since this type of wording just takes up space and adds little meaning to the title. Another consideration to keep in mind when you are creating a title is that if this document is eventually published, it should contain the proper keywords so that it can be found when someone does an electronic search for a specific document. If you think other individuals might use specific keywords to try to retrieve a document such as the one you produced, make sure these keywords are in its title.

If you are working in an engineering firm, the client name and full business address are given a heading of *Prepared for:* above the information. If you are a student submitting a report, put your instructor's name after this heading. Sometimes your instructor will also want you to indicate the course and section number in this part of the title page. Your name and your complete business address should be preceded by a heading of *Prepared by:* If multiple authors wrote the report, include all names at this point. The submission date is important to establish the paper trail and to document your work.

Mechanics. In most cases, all text on the title page should be centered between the left and right margins and between the top and bottom of the page. The font size for the title should be larger (about 18 point) than it is for the rest of the information on the title page. You may choose to give the headings (Prepared for, Prepared by, and Date submitted) a different text treatment (e.g., italics), although doing so is not required.

Example

Figure 5.4 shows an example of a title page prepared for an experimental report (discussed in Chapter 6) similar to what you might submit as a student. A title page for a professional report would probably contain most of the same information but would also include contact information and addresses for the person writing the report as well as for the person to whom the report is submitted.

5.6   Table of Contents

Function. A table of contents enables readers to find a specific part of your report quickly and easily. If your report is 50 pages long, imagine the difficulty the reader would have if he had to thumb through the entire document to find what he needed. You are probably familiar with a table of contents because each of your textbooks likely has one.

Form. The table of contents should list all the major headings and most of the subheadings in your document, depending on the number of subheading levels you have. If you used just one or two levels of subheadings, they should generally be included in the table of contents. If you used more than two, you may omit the third and fourth levels if they clutter the table of contents. Appendixes should also be listed, by letter (or number), in the table of contents. (Appendixes are described in more detail subsequently.)

Determination of Stress and Strain Concentration

Prepared for:
John B. Ligon
Professor of Mechanical Engineering

Prepared by:
Claire E. Hoffman
Graduate Student

Date submitted: April 23, 2005

Figure 5.4. Example title page for student experimental report

Mechanics. The major headings should be located along the left margin of the page. Subheadings should be indented by a small amount beneath their appropriate major headings. The page numbers should be included along the right margin of the page. You may want to include *leaders* (a dotted line) between the heading title and the page number for clarity. Leaders enable the reader to determine quickly the page number she is seeking.

Example

Figure 5.5 shows an example of a table of contents for a technical report. Note the use of second- and third-level headings in the example. Note also the extra space between major headings, which sets them off from one another.

Table of Contents

Figure 5.5. Example table of contents

5.7   List of Figures

Function. Because graphical communication is an important part of technical communication, most reports will include several figures. This fact is especially true for reports that describe significant experimental work (such as a graduate thesis or dissertation). If this is the case, a list of the figures generally accompanies the report so that individual figures can be quickly and easily found. In some cases, you may also need to include a list of tables in your document, especially if you have several tables with a great deal of information in them.

Form. The list of figures is set up much as a table of contents; however, figure numbers and labels are listed instead of document section numbers. The list of figures is usually placed directly after the table of contents. A list of tables would typically follow the list of figures.

Mechanics. As with a table of contents, the page number for a given figure is located at the right margin of the page. Once again, leaders enable the reader to find the page number that corresponds to a particular figure quickly and easily.

Example

Figure 5.6 shows an example of a list of figures that might be found in a technical document.

5.8   Appendixes

Function. Appendixes are often included in a report and typically contain large amounts of data and other supporting information. This supporting information needs to be included somewhere in the report but should not be part of the main body. Readers do not want to go through all the details when they are reading the report, but they may want to refer to the details in an appendix at some point for clarification.

Form. Appendixes are included at the end of the document, usually after the references. The type of information that might be included in an appendix is as follows:

■ Raw data
■ Design calculations
■ Source code for a program
■ Photographs of an experimental setup or of specimen failures
■ Graphs and charts not needed in the body of the report but possibly of interest to the reader

---

List of Figures

Figure 5.6. Example list of figures

Mechanics. Each appendix should start on a new page and be located at the end of the document. If you are compiling a large report that will be put into a three-ring (or similar) binder, you may want to put tabs on the first page of each appendix so that the reader can quickly locate each of them.

Each appendix should have a label and a title associated with it (e.g., Appendix A—Source Code for C++ Program, Appendix IV—VOC Reduction Data), and the labels should be either letters of the alphabet (preferred) or numbers, usually Roman numerals. Each appendix should be referenced somewhere in the body of the report. For example, appendix referencing should follow this style:

> Figure 3 shows the results from a typical tension test. See Appendix B for results from all tension tests conducted.

The appendixes should be labeled in the order in which they are referenced in the document (i.e., the first referenced appendix should be Appendix A, the second referenced appendix should be Appendix B, and so on). As mentioned previously, *each* appendix should be referenced somewhere in the body of the document. If an appendix is not referenced, do not include it in the final report.

Appendix B—Tension Test Results

Table B.1: Tension Test Results for 5/16″ Hole Diameter Base Clamped with Steel Plate

Load Angle Degrees	Test	Failure Mode	σ_u (ksi)	P_u (kip)	σ_y (ksi)	P_y (kip)	$\sigma_{4\%}$ (ksi)	$P_{4\%}$ (kip)
0	A	Bearing	43.5	5.1	38.3	4.5	7.7	0.90
0	B	Bearing	43.9	5.1	40.1	4.7	7.4	0.86
0	C	Bearing	44.6	5.2	37.4	4.4	7.4	0.87
0	D	Bearing	43.3	5.1	38.1	4.5	8.0	0.93
22.5	A	Flexural	28.1	3.3	28.1	3.3	5.7	0.67
22.5	B	Flexural	28.5	3.3	28.5	3.3	5.5	0.65
22.5	C	Flexural	27.2	3.2	27.2	3.2	5.7	0.67
22.5	D	Flexural	30.1	3.5	25.2	3.0	5.6	0.65
45	A	Flexural	23.8	2.8	23.3	2.7	4.9	0.57
45	B	Flexural	23.2	2.7	23.3	2.7	4.7	0.55
45	C	Flexural	24.1	2.8	23.7	2.8	4.3	0.50
45	D	Flexural	27.2	3.2	27.2	3.2	4.5	0.53
67.5	A	Flexural	28.0	3.3	28.0	3.3	4.7	0.55
67.5	B	Flexural	26.7	3.1	25.7	3.0	4.5	0.53
67.5	C	Flexural	30.1	3.5	30.0	3.5	4.6	0.54
67.5	D	Flexural	31.3	3.7	31.3	3.7	5.0	0.58
90	A	Flexural	34.6	4.1	34.0	4.0	4.4	0.52
90	B	Flexural	34.1	4.0	34.0	4.0	4.4	0.52
90	C	Flexural	28.5	3.3	27.0	3.2	4.2	0.49
90	D	Flexural	33.6	3.9	31.9	3.7	4.5	0.53

Figure 5.7. **Example appendix**

Example
Figure 5.7 shows an example of a short appendix that might be included in a technical report. In reality, most appendixes are typically several pages long.

5.9 Document Files

Function. One "headache" you will have to deal with when you are using a computer for technical communication is working with electronic files. Although computer systems have been standardized somewhat, you must still be aware of some potential issues, especially if you are working on a collaborative project with a colleague. Document files are used to store your document. If you want to work on a writing project collaboratively, the document file can be exchanged between authors, with each contributing his part and then passing the file on to the next author.

Form. Document files contain the text of your work. In most cases, document files can also contain equations, figures, and tables embedded within them. Several types of document files are used, and each is useful for specific applications. The basic document file types are discussed next.

Text files. A *text file* is the simplest type of document file; however, such files are generally platform dependent, which means text files created on a Mac will often appear differently on a PC. This type of file contains only text—no figures or equations. Word-processing software may embed special characters, such as carriage returns, within a text file as it is writing the text to the file. Take care that these characters are not included in your text file by specifying the correct options when saving the file. A text file is useful for exchanging information from one type of software package to another. For example, you could write a C++ program in a word processor and write it out to a text file. The contents of the text file could then be compiled and the program executed. In other cases, you might use a data acquisition system that writes your data to a text file. This text file could then be brought into a document for editing and cleanup so that the data become integrated within the document. Text files usually have the file extension *.txt* so that their file type can be readily determined.

Native files. A *native file* is the type of file that is automatically written by your specific word-processing package. For example, Microsoft® Word will automatically write a .doc file, StarOffice™ will automatically write a .sdw file, Word-Perfect® will automatically write a .wpd file, PageMaker® will automatically write a .pm6 file, and FrameMaker® will automatically write a .fm file. Native files will contain all the customized formatting, embedded figures and equations, and other special characters you used in creating the document. If you are working on a writing project, use only one type of file so that all your customization is maintained. Most word processors have some sort of document translator for reading from or writing into files that are not in their native format. However, although these translators are improving, they will *rarely* do a perfect job of translation when migrating a document from one file type to a different one. Another frequent problem with working with native file types occurs when a person assumes a colleague is using the same word processor. If that person then sends a file in native format to her colleague, he may or may not be able to view it correctly, depending on software availability. For example, one of the authors of this text almost always uses FrameMaker for word processing. Colleagues who use Microsoft Word often erroneously assume she is also using Word and send her Word documents. Not only is receiving such files a source of irritation, but it is also inefficient. Although FrameMaker can read and write a Word document, the formatting rarely comes through perfectly, which means files must be resent or meanings guessed. If you want to send someone a file and you are not sure which word processor the person has access to, send your file either in a PDF format or in an RTF format (discussed subsequently) to ensure its ultimate readability by the recipient.

Portable document files. The portable document file (PDF), designated by the extension *.pdf,* is recognized as a standard for exchanging documents while maintaining all formatting and special characters. This file type was developed by Adobe® but is widely used. This type of file is meant to be a medium for exchange that is independent of the native word processor in which it was created. Many word processors can print to a PDF file, but if not, a number of distillers are available for the creation of a PDF file from virtually any type of native document. In most cases, a PDF file is created as a read-only file and is therefore not a useful format for exchanging files for collaborative work. If, however, you want to send a file to an individual merely for reference or for viewing, this format is probably the best one to use. One of the most common uses for a PDF file is the creation of a document that is accessed through a Web site. If a person clicks on a Web page link that represents a PDF file, the viewing software is launched and she is then able to see or print the document in its correct format. The PDF files created for Web sites can also be set up so that they are nonprintable. This option is useful if you want someone to be able to read your document but not be able to print it. For example, your professor might put examples of well-written memos on a Web site as PDF files but make them nonprintable so that a student couldn't print one out to turn in for an assignment.

Rich text format files. A rich text file (RTF), designated by the extension *.rtf,* is used for the exchange of documents between different word-processing packages. The format for this type of file is standardized across the industry, so most formatting and special characters will be translated accurately. Most times, RTF files are fairly transportable between packages; however, some problems may occur with figures and equations, depending on the robustness of the RTF translator embedded in your word processor. One problem with an RTF file is that it is much larger than the same document written in a native file type. This increase in size is a small price to pay, however, if the use of an RTF file can save you a significant amount of time in reformatting work. Once you have read an RTF file into your word processor, you can usually write it out into native file format for subsequent editing if you choose.

Mechanics. Most word-processing packages automatically supply the standard file extension for the type of file you are writing (e.g., .doc, .sdw, .rtf, .pdf). If not, be sure to supply the correct extension so that you know what type of file you have later. If you are working collaboratively on a writing project, make sure you all agree beforehand on the specific word processor you will be using. Doing so will help prevent unnecessary formatting problems as you proceed with your work.

5.10   Web Pages

Function. Web pages are online documents you can use to convey information to others. Unlike a report or memo you give your boss, who may be the only other person to read it, documents posted on the Web may be widely viewed and read.

In some cases, you might use a Web site as a repository for data files that can be easily shared among colleagues. Rather than e-mailing files to others, you can put files on a Web page for team members to download as needed. In this case, you should probably use a *secure* Web site for the data transfer, unless the information is not sensitive and you don't mind if others outside your group have access to it.

Form. Web sites can take many forms; however, for technical documentation, simpler is probably better. When designing a Web page, you should think more about its *structure* than its exact appearance. Web pages will often appear differently on various computer platforms and with various browsers, so spending a great deal of time on formatting is probably not productive. Remember that not everyone who views your page will be using the same computer system as yours, so you should plan for this uncertainty and design the page for what you think they might have. Avoid the overuse of animation: not everyone who wants to view your page has the most up-to-date system, and some people may not have the fastest Internet connections.

A Web page should be structured so that readers can quickly find the information they seek. Viewers do not want to read through vast quantities of text to find pertinent information. They want to be able to glance at a page, find the link they need, and proceed to the next page, and so on, until they finally reach their destination. Imagine the frustration viewers would experience if they wanted to place an order from an online merchant and were required to scroll through an entire long list of available items before they could find the one they wanted. Instead, the merchant would be wise to categorize his merchandise so that viewers could search by type of product, which would reduce the quantity of information they would have to sort through before finding the desired product.

Some basic rules to follow when you are creating a Web page are as follows:

- The purpose of the Web page and the information contained thereon should be readily apparent at first glance.
- Your e-mail address should be on the page so that viewers can reach you with questions if necessary. You should be aware that some spiders traverse the Web searching for e-mail addresses on Web pages to add to spam lists. To avoid this, include your e-mail address in a different form or insert spaces in it so it is not a clickable link (e.g., mailid @ server . com).
- If your site has several layers, make sure viewers can easily navigate between layers and back to the beginning if necessary.
- If your page is part of a larger page, make it easy to find from the main home page and make navigation back to the home page relatively easy.
- Most Web pages have embedded graphics. Remember that not all users have the same computer power you might have. Too many graphics or files with resolutions set unnecessarily high will cause users to become frustrated with the amount of time required to load your page. Use

graphics only to enhance the readability or the attractiveness of your page and don't overdo it.

■ Keep the amount of information on a given single page of your Web site to a minimum; however, make sure you provide the viewers with enough information to answer their questions.

■ Include links to other Web pages that might interest your viewers.

■ Periodically review your Web page to make sure it is still current and the information on it is still accurate.

■ Include a site map if your page contains many linked documents so that viewers can quickly go to the specific page they want.

Mechanics. Web pages are most often created, or *written,* by using *hypertext markup language (HTML).* Many modern-day word processors can write HTML documents directly; however, this capability should be used only for the most rudimentary documents. Complex pages are best created strictly through HTML or a software package specifically designed to help create Web pages and not a word processor. Although all the intricacies of HTML are beyond the scope of this text, a few basic HTML *tags,* or commands, and pointers are worthy of mention.

Text within an HTML document is tagged so that when a person accesses your Web page, her browser will read the file and interpret the tags according to its default settings. Thus, tagging text as a first-level heading in your HTML document is preferable to specifying it as 18-point Arial Bold (the reader's browser may not display the text this way).

Tags are included both before and after a string of text: in essence, you turn a feature on with the first tag and off with the second tag. Tags are enclosed within angle brackets (<>) in your document, and the second tag usually includes a slash (/) to indicate you are turning the feature off. (*Note:* Some HTML tags are case sensitive, which means you must include them in the proper case, either upper or lower, to ensure they are interpreted correctly.)

For example, suppose the following line of text exists in an HTML document:

<B><I>Sample HTML sentence.</I></B>

The <B> and the <I> are the HTML tags for boldface and italics, respectively. If this sentence is then interpreted by a browser for display, it will appear as follows:

Sample HTML sentence.

Note that if the sample sentence were in your HTML file as

<B>Sample <I>HTML</I> sentence.</B>

Its appearance interpreted through the browser would change to this:

Sample *HTML* sentence.

In this case, only the *HTML* has been tagged for italics, and the entire sentence has been tagged for boldface.

HTML tags are also included to create ordered (numbered) or unordered (bulleted) lists, to signify titles or headings, to center blocks of text within a page, and so on. The <HTML> tag must be used to enclose an entire HTML document. This tag is used by the browser to determine where the file interpretation starts and stops. For example, following is a valid HTML file:

```
<HTML>
<HEAD>
<B>The Engineering Process</B>
</HEAD>
<BODY>
```

The engineering problem-solving process is a time-tested method based on logic and theoretical understanding. The following steps are often used in the engineering problem-solving process:

```
<OL>
<LI>Recognize and understand the problem.
<LI>Accumulate data and verify their accuracy.
<LI>Select the appropriate theory or principle.
<LI>Make necessary assumptions.
<LI>Solve the problem.
<LI>Verify and check the results.
</OL>
</BODY>
</HTML>
```

Your browser would interpret the preceding HTML file as follows:

The Engineering Process

The engineering problem-solving process is a time-tested method based on logic and theoretical understanding. The following steps are often used in the engineering problem-solving process:

1. Recognize and understand the problem.
2. Accumulate data and verify their accuracy.
3. Select the appropriate theory or principle.
4. Make necessary assumptions.
5. Solve the problem.
6. Verify and check the results.

Literally hundreds of tags are available for you to use to create your Web page. If you are interested in learning more about HTML tagging and use, first, you can locate and study one or more of the numerous texts written about HTML (e.g., Ernst and Engst 1995; Williams and Tollett 1998). Second, you can use your browser to view the source code for any page you are currently displaying. In this way, you can see how the HTML tags are used within a file and then how your browser interprets those tags. Third, you can consult the following Web site: http://archive.ncsa.uiuc.edu/General/Internet/WWW/HTMLPrimer.html

5.11 Exercises

1. Create a style sheet that shows 10 combinations of font and size.
2. Create a title page for a report you wrote in one of your classes.
3. Go to your university home page and spend some time browsing. Write an e-mail to your instructor describing what you think are the best and worst features of the site.
4. As a group, examine the Web site of a company that makes consumer products. Write a group memo to your instructor that covers the following points:

   ■ The most attractive aspects of the site
   ■ The least attractive aspects of the site
   ■ The ease of navigation on the site
   ■ The overal effectiveness of the site

   Print the home page to turn in with your memo.
5. View the Web site of a significant professional society in your field. In a memo to your instructor, describe the major types of information found on the site. Determine what type of student involvement is available through the society and include this information in your memo.
6. Make a checklist that can be used to develop a Web site. One item might be as follows: Did you include links to other related Web sites?

5.12 References

Ernst, T., and A. Engst. 1995. *Create your own home page.* Indianapolis, IN: Hayden Books.

Williams, R., and J. Tollett. 1998. *The non-designer's Web book: An easy guide to creating, designing, and posting your own Web site.* Berkeley, CA: Peachpit Press.

Technical Written Communication

Writing is an activity engineers engage in virtually every day, from writing e-mails to composing formal technical reports. As such, you must understand the various forms of written communication you might be expected to produce, critique, or evaluate. Writing, especially technical writing, is a skill best developed through continuous practice. Therefore, you must strive to improve your writing skills and to practice them regularly. At first, writing may seem to be painstaking, with the blank piece of paper staring you in the face for a long time before you begin to put your thoughts down on it. However, with practice and experience, you will be able to write increasingly well. In this chapter, we describe many of the forms of written communication you might use as an engineering student or as a professional. Although a given situation may call for a type of documentation not presented in this chapter, the more common forms of technical written communication are described.

6.1 Summarizing Information

Many times in the engineering field or class, you will be asked to summarize your findings in a brief, to-the-point document. The length of this document may vary depending on the context. When writing this type of document, you should choose each word carefully and structure your sentences to minimize wordiness. The two main types of documents you will write that summarize information are an executive summary and an abstract. In addition, many times you will be asked to supply three or four keywords or phrases that summarize the contents of a document. Supplying relevant keywords and phrases is likely to become increasingly important as Internet search engines are used more often to look for certain words or phrases.

6.1.1 Executive Summary

Function. An *executive summary* gives a complete overview of a report. It is an independent element of the report and, thus, should be able to stand alone. The executive summary is important because it is often the only part of the report that is read, which is frequently the case with managers, who may have several engineers or other staff members reporting to them. A manager in this position wants to be able to learn the main points of a report quickly and does not want to read 20 pages of text. If any specific questions are raised while the manager is reading the executive summary, she can look through the entire report to find the necessary details. Learning to write effective executive summaries will likely serve you well as you move into the corporate world, where time is critical. Executive summaries should be written with all audiences in mind and should be concise because they are commonly used for in-house technical reports as well as for reports distributed to the clients of a firm.

Form. An executive summary typically consists of three distinct parts:

1. A summary of the key points of the introduction, objectives, background material, procedure, and results of the report
2. Conclusions
3. Recommendations

In many exective summaries, these parts are often separate sections and each has a heading: Summary, Conclusions, and Recommendations. Such subdivision allows a reader to find the information he requires without having to read the entire exective summary.

Mechanics. An executive summary is usually about one typewritten page, or about 500 words, long. In some companies, a limit of two pages is put on the executive summary, and in others, the limit is a percentage of the report length (e.g., 5 to 10 percent). The executive summary can be a challenge to write because it must be a clear and concise statement of the three essential elements (discussed in the preceding section). When writing an executive summary, you must include introductory material to place the report in context for the reader, describe the significance and implications of the results, and clearly state your conclusions and recommendations. You should write the executive summary after you complete the entire report, and you should make sure all the key points in the report are present in the executive summary.

Design of a Low Cost Portable Combustion Data Analysis System

Executive Summary

Summary

Current federal air-quality standards require all new automobile models to pass emissions tests before they can be mass-produced. Government testing is conducted by determining tailpipe emissions; however, measuring tailpipe emissions alone is not efficient during the development phase of a new vehicle model. Testing in a dynamometer chamber is the accepted method for conducting the emissions tests during the vehicle design; however, it is costly and not efficient for testing during the development phase of a new automobile model. In this project, a low-cost alternative to dynamometer chamber testing was developed. This development could potentially translate into a significant reduction in the cost associated with the design of a new automobile.

Data were gathered through several pressure and temperature sensors mounted on the cylinders, the exhaust manifold, and the catalytic converters of the test vehicle: a sport-utility vehicle. These data were input into a data acquisition board and the signals were conditioned and converted from analog to digital and used to approximate total emissions through several computational routines. The data analysis system that was developed uses a LabVIEW program for sampling data, a C++ program for performing computational work, and a Visual Basic macro for importing the data into an Excel spreadsheet. Once the data have been imported into Excel in a usable format, they can be further manipulated and output to graphs as necessary.

Conclusions

Through this project, a data analysis system for estimating nitrous oxide (NO) emissions from vehicle cylinders was successfully developed. This system enables accurate prediction of emissions without the expense of a dynamometer chamber. The system uses a simple Windows NT PC platform for analysis, and standard sensors and software for data acquisition. The system is cost effective and user friendly, and it should greatly reduce the time and expense involved in the design of new vehicle models.

Recommendations

Future enhancements of this data analysis system would be to improve the user interface for performing the analysis and to streamline the system into a single routine. Another enhancement would be to migrate the system to the MTS/CAS system so that additional data collected throughout the vehicle could be incorporated into the analysis.

Figure 6.1. Example executive summary

Example

Figure 6.1 shows an executive summary derived from the design project report shown in Figure 6.5. Note that the executive summary in Figure 6.1 is separated into sections that include a summary, conclusions, and recommendations. Figure 6.2 shows a poor executive summary. It is not separated into sections. Although this example is not a good executive summary, it is an acceptable *abstract*.

6.1.2 Abstract

Function. The function of an abstract is similar to that of an executive summary. An *abstract* summarizes the content of a technical *paper* (an executive summary

---

Design of a Low-Cost Portable Combustion Data Analysis System

Executive Summary

Current federal air-quality standards require all new automobile models to pass emissions tests before they can be mass-produced. Government testing is conducted by determining tailpipe emissions; however, measuring tailpipe emissions alone is not efficient during the development phase of a new vehicle model. Testing in a dynamometer chamber is the accepted method for conducting the emissions tests during the vehicle design; however, it is costly and not efficient for testing during the development phase of a new automobile model. In this project, a low-cost alternative to dynamometer chamber testing was developed. This development could potentially translate into a significant reduction in the cost associated with the design of a new automobile. Data were gathered through several pressure and temperature sensors mounted on the cylinders, the exhaust manifold, and the catalytic converters of the test vehicle: a sport-utility vehicle. These data were input into a data acquisition board and the signals were conditioned and converted from analog to digital and used to approximate total emissions through several computational routines. The data analysis system that was developed uses a LabVIEW program for sampling data, a C++ program for performing computational work, and a Visual Basic macro for importing the data into an Excel spreadsheet. Once the data have been imported into Excel in a usable format, they can be further manipulated and output to graphs as necessary.

Figure 6.2. Example of a poor executive summary

typically summarizes a technical *report*) and is usually no longer than a single paragraph. The function of an abstract is to inform the reader of the contents of and key points in a paper. Thus, an abstract allows a reader to determine quickly whether reading the entire paper would be worthwhile. This function is different from the function of an executive summary in that an executive summary is intended to potentially replace a report for people who do not have time to wade through all the details in the report. In contrast, an abstract summarizes the main ideas found in a paper, but an abstract will not replace the need to read the paper.

Sometimes a distinction is made between an *informative abstract* and a *descriptive abstract.* In the preceding paragraph, the definition of an abstract is that for a descriptive abstract. An informative abstract is analogous to an executive summary. A report can have both an abstract and an executive summary. In such a case, the abstract is a short description of the report that allows the reader to decide whether to continue on to the executive summary.

Form. The basic form of an abstract is the same as that of an executive summary, except separate headings are not used and an abstract is usually shorter. You should make sure your abstract contains pertinent background information, the results achieved, and conclusions. Recommendations for future work may or may not be included in an abstract.

Mechanics. Abstracts are typically no longer than a single paragraph and may be limited to as few as 100 to 150 words. Separate headings are not included in an abstract. In other words, an abstract is like a preview to a movie: readers will determine from the content of an abstract whether they want to pay the "price of admission" to read the entire paper. The abstract should be written after the document is completed. In many cases, the font used for the abstract differs from that used for the remainder of the document, which sets it off from the main body of the paper.

Example

Figure 6.3 shows an example of an abstract that is included with a technical paper. (*Note:* The paper is shown in its entirety in Figure 6.7.) Note that the abstract includes what was done, why it was done, and what the results were.

6.1.3 Keywords

Function. *Keywords* are the briefest possible means of summarizing a document. A keyword can consist of a single word, or it can consist of a group of two or three words. When providing keywords for a technical paper or report, you should usually supply three or four keywords (or phrases) that highlight its contents. Good keywords will likely allow your document to be found during an online search of a specific topic, whereas poor keywords will not.

Another common use of keywords is found in documents such as résumés (described in Chapter 10). Recruiters search many online résumé databases, using specific keywords, so you must keep in mind the standard keywords that will ensure your résumé is among those found during this type of online search.

Form. As mentioned, keywords are typically two or three words that "go together." Examples of keywords include *artificial blood, Carnot cycles, diesel engines, prestressed concrete, structural engineering, thermodynamics,* and *unit operations.*

Mechanics. When selecting the keywords that describe your paper, think, first, about the three or four key topics in it. Second, think about the standard phrases another person is likely to use to search for information on a topic related to the

---

Abstract. Wood-peg-connected timber frames require the use of an analysis procedure that accounts for the connection behavior. The peg stiffness can be modeled by using short elements whose axial stiffness is determined by assuming that the peg acts as a simply supported beam with a concentrated load at midspan. The concentrated load is the tenon bearing on the peg, and the supports are the sides of the mortise. This approach has been shown to work when its results have been compared to test data on the behavior of a single-connection design. This modeling approach has been extended to a fairly large frame to examine the sensitivity of the frame to various modeling assumptions, particularly the effects of member-to-member contact at a joint. If the effects of contact are not accounted for, estimations of some load effects may be underestimated by more than 100 percent, although most underestimations are 25 percent or less.

Figure 6.3. Example abstract

information presented in your paper. Third, make sure your keywords are as specific as possible so that your document is found as a person's search is narrowed. For example, instead of a keyword such as *pavement*, a keyword of *bituminous asphalt pavement* would be much more specific and descriptive of the contents of a given paper. When possible, use standard keywords (Library of Congress, 2004) or those suggested by the group to which you are submitting your paper. Keywords are typically placed immediately after the abstract (see Figure 6.7). However, you may be required to place them in another location, such as on a separate sheet, so be prepared to follow your instructor's or the publisher's guidelines.

Examples

For the abstract shown in Figure 6.3, the following keywords were selected to describe the content of the paper:

Keywords: Mortise and tenon, timber engineering, heavy timber, wood engineering.

6.2 Proposing Ideas

You will likely be required at some time to develop a bid to (a) obtain a project, (b) develop a plan of action for a new line of inquiry, or (c) obtain funds for a new project in research and development. The type of documentation you will create to suit this purpose is called a *proposal*. A proposal has an audience of one or more persons who will decide whether you, your team, or your company gets the work you propose to do. The proposal can be *solicited*—an agency requests the proposal, usually through a *request for proposal (RFP)*—or *unsolicited*—you submit the proposal without being requested to do so. Furthermore, the proposal can be *internal*—directed to a person inside your company—or *external*—directed to an outside agency. Whichever type of proposal it is, you are trying to convince the reader that you, your team, or your company is best able to do the job. Thus, you need to persuade the proposal reader that your approach to the work is the best way (in terms of both cost and effectiveness) and that the people who will be doing the work are the best ones to do it.

6.2.1 Internal and External Proposals

Function. A *proposal* is a detailed plan for investigation of a design or process alternative or for a research project, or it is a request to do an engineering project. Proposals can contain provisions for conducting experimental work as well as the development of basic theory. Before you write a proposal, you should have a well-defined plan of action that is likely to bring about the desired results. The writing style for a proposal is similar to that used in other forms of written technical communication; however, using first person in a proposal is often acceptable.

Form. The form of a proposal could be a memo, for an internal proposal, or a formal proposal, typically for external proposals. A proposal is usually divided into a number of subsections, as outlined next. An internal memo proposal will likely include all the following information but will do so in a condensed manner, whereas a more formal proposal will include specific sections, of which the following are typical.

Project summary. Similar to an abstract for a paper, a Project Summary section is often included at the beginning of a proposal to define, upfront, the type of work you intend to carry out.

Introduction. In the Introduction, you clearly define the problem and the need for the investigation. You should describe the specific reasons why the design alternative, process alternative, or research project will be beneficial to your company, your client, the funding agency, society, engineers, or whoever might benefit from the work.

Results from previous work, or background. If other engineers have investigated a similar problem, you should present their results in the Background section. Typically, this section is a compilation of the results from previous work as reported in books and journals. As for technical reports, you will likely need to go to the library or look through company archives to search for the information to be included in this section of the proposal. Whenever you mention others' work, make sure you cite the source of your information properly. (Citations are covered in Chapter 3.)

Objectives. In the Objectives section of the proposal, you describe the project goals and the outcomes you expect to produce by carrying out the investigation. Specific goals and objectives should be outlined, along with an indication of how you will measure success.

Significance of proposed work. Although information related to the significance of the research or development is usually included in the Introduction, sometimes including a specific section dealing with it is beneficial. This section, usually only one or two paragraphs long, emphasizes why the proposed work is important to your company or the funding agency to which you will submit the proposal. Again, the importance is related to the benefits that will accrue to the various groups mentioned in the Introduction. The description of the benefits of the work is vital to a successful proposal. No matter how well the remainder of your proposal is written, if the proposed work does not appear valuable, no one will want to spend time or money performing the work.

Project description. In the Project Description section, you outline the experimental, numerical, or theoretical methods you will use to conduct the investigation. You also describe the equipment and software that will be required for the proposed work. A project time line and the specific tasks to be carried out during the investigation should be mentioned as well.

Future work. The Future Work section is optional. If included, it should outline where you think the proposed work will lead—that is, what you think the next step in the investigation will be after the current project is completed.

Personnel and budget. In the Personnel and Budget section, you detail who will conduct the investigation, their qualifications, and how many person-hours will be required for the work. You also outline the cost of purchasing equipment or supplies for the proposed project. If travel will be required, estimates of the cost should be included in the budget. In some cases, you may need to subcontract a portion of the research or development. If so, this cost must also be included in the budget.

References. In the References section, list all the previous work (e.g., journal articles and books) cited in the proposal.

Appendixes. You may need to include information in one or more appendixes if, for instance, the RFP you are responding to requires it, or you may want to supplement the body of the report with information in an appendix.

Mechanics. As for other forms of written technical communication, use a font size and treatment that will make headings stand out from the remainder of the text. The use of first person is often considered acceptable in proposals. However, even if it is, it should be used sparingly. First person is most appropriate when you are describing what you or your team plan to do. Use graphics as appropriate throughout your proposal. A Gantt chart (discussed in Chapter 9) is particularly useful for graphically showing a proposed project time line.

Example
Figure 6.4 shows an example of a proposal.

6.2.2 Proposals Written in Response to an RFP

As discussed previously, one type of proposal you may need to write is an *external, solicited* proposal, most of which will be written in response to an RFP. The *function* of this type of proposal is the same as for any proposal: you want to get the work. The *form* of this type of proposal depends entirely on the RFP. When writing a proposal in response to an RFP, you must compose your proposal in the *exact* form requested by the funding agency or company. If the RFP does not specify a form, use a logical format similar to that just discussed. Furthermore, you must address all the *specific* issues the RFP indicates should be addressed. Sometimes you can address more than these issues, but make sure the RFP allows you to do so. Often, if you do not use the correct format or neglect to address a specific issue, your proposal will be rejected immediately and without further review. The key to responding to RFPs is to do *exactly* what the RFP requests. One additional, vital point is to ensure the proposal arrives at the funding agency by the deadline.

**Proposed Integrated Approach for the Electromagnetic
Characterization of the Automobile Environment**
F. W. Perger and R. S. Rousselle

———— Title and Authors

Project Summary

The explosion of mobile communication in recent years has resulted in a corresponding increase in interest for improved performance from antenna structures for automobiles. However, the automobile antenna must also meet mechanical and aerodynamic constraints, which suggests a design approach that integrates electrical and mechanical software and modeling. An approach has been developed that promises to accomplish the task of designing new antenna structures through scientific visualization that takes into account the automobile environment. The proposed research would explore the feasibility of applying this approach to realistic automobile examples.

———— Brief Overview of Proposed Project

Introduction

The idea of sharing design information from or with other engineering disciplines is often considered "untraditional," especially in the field of electromagnetics (EM), because of its uniqueness and the degree of specialization required. The advent of computer workstations, however, has bridged this gap in the sense that computer modeling, design, and analysis techniques are similar across engineering disciplines. Taking advantage of these parallels allows an efficient computer aided engineering (CAE) environment incorporating EM to be realized. One problem faced in EM analysis methods is a lack of tools that can be used to visualize the results from numerical processors such as NEC2 (Burke and Poggio 98). Scientific visualization is a complicated subject since the data representation is often unique to a given application. However, through the juxtaposition of visual, textual, and numerical forms, a visualization system can make sense out of masses of otherwise incomprehensible data (Arrott and Latta 63).

———— Problem Definition

———— Reference Cited Using Author/Page Number Method

Background

The traditional engineering design process is essentially a 1-D approach often referred to as *over-the-wall engineering.* As the name suggests, the design is passed from one rank of engineering discipline to another. By the time a design reaches the manufacturing state, it is often too late to make minor changes that could have resulted in significant costs savings because no efficient way to compromise on design criteria was available among the various engineering disciplines.

———— Problem Background and Recommendations from Previous Researchers

To increase the efficiency of the design process, many companies are adopting a philosophy of concurrent team engineering. In this process, designers, analysts, and manufacturers work together as a team, throughout the design process, to develop a cost-effective, optimal solution. CAE software provides a common link among team members and is essential to this design philosophy. Through the CAE software, every team member has access to the design through all stages of the process.

Numerous recommendations for the development of EM design systems that include aspects of scientific visualization have been made (Miller 26; Miller, Kruger, and Moraites 127; Siarkiewicz 119). While reasonably accurate and reliable software packages for modeling of EM fields received by antennas exist, sophisticated visualization tools that could be used along with mechanical CAE tools to design antenna structures in the automobile industry are lacking.

Objectives

The primary goal of this project is to test a framework for analysis and design of an antenna structure for an automobile. The specific objectives of the project include the following:

———— Specific Goals and Results for Proposed Project

- To develop a realistic 3-D model of an automobile in CAE software that includes material characteristics relevant to EM analysis methods
- To modify the NEC2 program to enable it to receive geometric and material data from the CAE software for EM analysis
- To test the framework through an iterative design optimization that takes into account structural, aesthetic, and EM characteristics

Figure 6.4. Sample proposal

Significance of Proposed Work

The rapid growth of wireless communications technology has spawned a corresponding exponential increase in the need to communicate with and between automobiles. However, filling this need has been met with difficulties because of the many competing design constraints for automobiles and their antennas. The advent of wireless applications into the gigahertz bands, such as global positioning systems at 1.575 GHz, has driven the need for automobile antenna systems capable of meeting new EM requirements as well as structural, aerodynamic, and aesthetic design constraints. This need suggests the importance of the development of an *integrated* approach for the design of automobile antennas in ways that were unheard of until recently, and this integrated approach has now been made possible through advances in the processing and graphical capabilities of modern-day workstations.

The research and development proposed in this project will fill a need in the automotive industry by providing engineers across several disciplines the ability to work simultaneously on the design of automobile models that are structurally sound, are aesthetically pleasing, and have high-quality EM characteristics. The design and analysis tool that is developed will provide a means for an efficient framework for concurrent engineering in the automotive industry.

> Demonstrating the Need for the Project

Project Description

In the concurrent engineering environment, one of the largest hurdles to overcome is making the ideological jump from model aids to actual design tools. With the design tool described in this proposal, the engineer can proceed through the entire design process without changing software packages. Figure 1 shows an overview of this design process. With this software tool, every aspect of the system is developed by using the same computer model and a shared interest in the design criteria.

> Figure 1 Referenced in Text

Initial development of an integrated software tool for the automotive industry has been promising. Through the proposed project, this work will be expanded, refined, and fully tested. Specific steps to be accomplished during this project are as follows:

- The preliminary software tool already developed in the feasibility investigation will be further optimized and adapted specifically for automotive work.
- Scale-model tests will be performed in the anechoic chamber for verification of the software results.
- A manual will be written to provide future users with insight into and instructions on the use of the integrated software tool.

> Specific Tasks to Be Completed

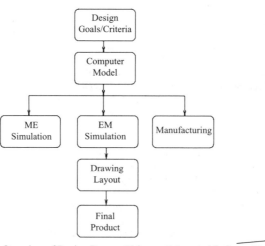

Figure 1. Overview of Design Process Using an Integrated Software Tool

> Figure 1 Labeled

Figure 6.4. Sample proposal (*continued*)

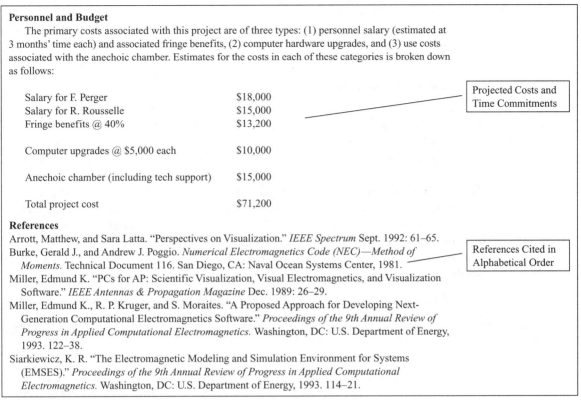

Personnel and Budget
   The primary costs associated with this project are of three types: (1) personnel salary (estimated at 3 months' time each) and associated fringe benefits, (2) computer hardware upgrades, and (3) use costs associated with the anechoic chamber. Estimates for the costs in each of these categories is broken down as follows:

Salary for F. Perger	$18,000	
Salary for R. Rousselle	$15,000	*Projected Costs and Time Commitments*
Fringe benefits @ 40%	$13,200	
Computer upgrades @ $5,000 each	$10,000	
Anechoic chamber (including tech support)	$15,000	
Total project cost	$71,200	

References
Arrott, Matthew, and Sara Latta. "Perspectives on Visualization." *IEEE Spectrum* Sept. 1992: 61–65.
Burke, Gerald J., and Andrew J. Poggio. *Numerical Electromagnetics Code (NEC)—Method of Moments*. Technical Document 116. San Diego, CA: Naval Ocean Systems Center, 1981.
Miller, Edmund K. "PCs for AP: Scientific Visualization, Visual Electromagnetics, and Visualization Software." *IEEE Antennas & Propagation Magazine* Dec. 1989: 26–29.
Miller, Edmund K., R. P. Kruger, and S. Moraites. "A Proposed Approach for Developing Next-Generation Computational Electromagnetics Software." *Proceedings of the 9th Annual Review of Progress in Applied Computational Electromagnetics*. Washington, DC: U.S. Department of Energy, 1993. 122–38.
Siarkiewicz, K. R. "The Electromagnetic Modeling and Simulation Environment for Systems (EMSES)." *Proceedings of the 9th Annual Review of Progress in Applied Computational Electromagnetics*. Washington, DC: U.S. Department of Energy, 1993. 114–21.

References Cited in Alphabetical Order

Figure 6.4. Sample proposal (*continued*)

Deadlines for RFPs are *true deadlines*; if your proposal is not at the agency on time, it is not accepted. In extremely rare instances, the funding agency may extend the deadline for *everyone,* but don't count on it. The *mechanics* for this type of proposal are the same as for any proposal, unless the funding agency says to do something specific that is different than discussed in this book. *Remember:* The funding agency is *always* right.

6.3 Reporting Information

As an engineering student or professional, you will often need to report the results of a study or an investigation you performed. Technical information is typically communicated as a report or as a paper. Two common types of reports are experimental reports and design reports. In the following sections, we describe these types of technical documentation. As for any type of documentation, you must have a clear understanding of the audience and the context before you begin writing.

Common elements found in most types of reports are a title page, a table of contents, a list of figures, and one or more appendixes. These elements are discussed in Chapter 5.

6.3.1 Technical Reports

Function. Technical reports are a crucial component of engineering documentation. *Technical reports* include design reports, analysis reports, and other engineering reports. Each gives an overall view of, as well as many of the details about, how a design, an analysis, or another engineering study was accomplished.

Form. When writing technical reports, you must write in a clear, concise style. Generally, the passive voice is preferred, and you shouldn't use first person. However, in some cases, you can use both the active voice and first person, especially if your report will be read by individuals without a technical background. Again, you must know your audience. A technical report is typically divided into several sections, as outlined next.

Title and author. Frequently, the title and author (or authors) are included on a separate title page, but a separate page is not always necessary. The title should help people decide whether they want to continue reading the report. The date of the report should also be included.

Table of contents. If the paper is fairly long, include a table of contents so that readers can find sections quickly and easily. The table of contents is a guide to the report.

List of figures. Technical reports often include several figures. Supplying the figure names and page numbers in a separate section following the table of contents can be helpful.

List of tables. Technical reports may also include several tables of data. Supplying the table names and page numbers in a separate section following the list of figures can be helpful. Note that the list of tables and the list of figures are sometimes in reverse order.

Abstract or summary. A well-written technical report usually begins with a short, introductory abstract or summary mentioning the major results or findings discussed in detail in the report. Such an abstract is an informative abstract. As mentioned in Section 6.1, the report may require both an abstract and an executive summary.

Introduction. The Introduction includes any pertinent background material and the objectives of the engineering study. It should also clearly state the problem being addressed. The background and the objectives may stand alone as separate sections of the report.

Proposed solutions. The Proposed Solutions section is not always included in a technical report, but it might be required, especially in design reports you prepare as college students. In this section, you discuss other possibilities that were considered and give the reasons why a particular system was chosen over a different one.

Main body. The bulk of the report goes in the main body. Note that this section won't be labeled "Main Body" but will have sections labeled according to the *specific* engineering solution you are documenting. In this section, you outline the engineering process as it was applied to the specific problem. The main body of the paper is usually divided into separate sections and subsections. The titles of the sections and subsections will depend on your specific design problem. One possibility is as follows:

- *Procedures.* In the Procedures section, you describe equipment, materials, software, analysis techniques, codes and standards, and other information required for the reader to understand how your system was designed or analyzed.
- *Results.* In the Results section, you describe your design or analysis in enough detail for the reader to understand exactly how the system will work. A *system* could be, for example, a chemical process, a building, a machine, or an electronic device.

Discussion. In the Discussion section, you discuss some of the important specific features of your design or analysis, any problems you encountered during the engineering process, and any unusual aspects of your design or analysis. You also discuss whether you believe your engineering choices were optimal. The reasoning in this section should be in-depth enough to support any conclusions and recommendations made in the next two sections. This section is sometimes combined with results in a Results and Discussion section or is sometimes omitted if discussion is integrated throughout the report.

Conclusions. The Conclusions section consists of an interpretation of the results. You must take a stand on the results and say what you believe the results imply. The Conclusions section should contain no information that hasn't been cited or discussed previously in the report. The conclusions should be based on the discussion of the results. Conclusions can be listed if desired.

Recommendations. In the Recommendations section, you call for specific actions based on the conclusions. This section may consist of a single sentence or be much longer. A list of recommendations is acceptable. This section is sometimes combined with the conclusions in a section titled "Conclusions and Recommendations."

References. You should list all references cited within the report. Most times, this list of references will be in a separate section at the end of the document.

Appendixes. Include appendixes with your report as needed to display data or results.

Mechanics. Headings for the various sections of a design report should clearly stand out from the text that follows or precedes each heading. Many times, headings will be in a slightly larger font size or different font style than that of the text of the report. Making headings stand out further through the use of features such as bold-face or italics is also acceptable. Some sections of your report will have subsections. Usually, the font and treatment (boldface or italics) of a subheading will differ slightly from that of the main headings so that the two are distinct. You may want to number each main heading for clarity; however, doing so is often unnecessary.

The first line of each new paragraph in a design report should be indented so that readers are aware of changes in topics or ideas. Alternatively, you can choose not to indent the first line of paragraphs but instead put a single blank line between two consecutive paragraphs. Paragraphs in design reports are sometimes double spaced rather than single spaced to allow readers the ability to make handwritten comments on the document as they are reading it.

The length of a design report will vary greatly depending on the specific project on which you are reporting. In general, documents 10 to 20 pages long are acceptable. A design report will be judged by the clarity of its ideas and expression, not on its bulk. As always in technical writing, clear, concise language is a must, and you should avoid excessive wordiness at all costs.

Example

Figure 6.5 shows an example design report. Note that this example does not have a table of contents, a list of figures, or a list of tables because of its short length. Note also that the main body of this design report has three sections: one labeled "Data Acquisition System," a second labeled "LabVIEW Interface," and a third labeled "C++ Program." In addition, the problem is stated clearly in a section labeled "Objective." Notice that a Discussion section is not included in this report.

6.3.2 Experimental Reports

Function. After you finish an experimental study or a series of experiments, you prepare a report outlining the data you gathered. This document is called an *experimental report.* As an engineering student, you will conduct laboratory experiments and then most likely will be required to submit experimental lab reports summarizing your results. As a practicing engineer, you will likely write experimental reports at some point in your career. An experimental report provides results that you obtained through your experimental work, outlines the procedure you followed so that someone else could conduct the same experiments later if necessary, and describes any problems or errors that may have occurred that could have affected your results.

**Design of a Low-Cost Hannah Hoffman, Jacob Duits, and Zachary Rossi
April 23, 2004 Portable Combustion Data Analysis System**

Summary

Current federal air-quality standards require all new automobile models to pass emissions tests before they can be mass-produced. Testing in a dynamometer chamber is the accepted method for conducting the emissions test; however, it is costly and not efficient for testing during the development phase of a new automobile model. In this project, a low-cost alternative to dynamometer chamber testing was developed. The data analysis system developed uses a LabVIEW program for sampling data, a C++ program for performing computational work, and a Visual Basic macro for importing the data into an Excel spreadsheet. This system can be used to perform emissions tests that use a standard Windows NT computer along with appropriate sensors, which results in a cost-effective, portable, easy-to-use system that greatly reduces the need for full-scale testing in a dynamometer chamber.

Results stated up front

Background

Passenger vehicles must pass emissions testing before the Federal government approves them for production. Government testing is conducted by measuring the emissions that exit the vehicle's tailpipe. During the model development phase, emissions are tested frequently so that problems can be quickly corrected; however, testing tailpipe emissions alone does not provide enough useful information to the design team. Since emissions problems can be caused at any of several fault points in the combustion system of an automobile, reducing overall tailpipe emissions usually means pin-pointing exactly where the problem is occurring along the combustion system.

Outline of problem to be solved

Detailed emissions data can be obtained through the use of a chassis dynamometer cell. Most automotive companies do not have a large number of these cells, and lead time for conducting a test in the cell is typically long. Further, the cells have a relatively high operating cost, of around $200,000 per session. The development of a low-cost alternative to testing in a dynamometer cell could potentially translate into a significant reduction in the cost associated with the development of a new automobile model.

Objective

The objective of this project was to design a portable system capable of acquiring and analyzing cold-start emissions data for 4-, 6-, 8-, 10-, and 12-cylinder engines. The data to be used in the analysis include that from cylinder pressure, spark timing, fuel injection, and emission analyzers. The system that was to be developed should enable output according to user-selected parameters, including both numerical and graphical formats.

Specific design criteria stated

Proposed Alternative

Early in the process, three potential design solutions were identified and investigated. The first proposed solution involved building an MTS/CAS system and using DSP Technologies hardware and software. Implementing this solution would require a number of new sensors and two DSP Technologies racks to be procured. New programs for the data acquisition and analysis would also be required. The cost to implement this solution would be in excess of $50,000 and the lead time for procurement of the DSP Technology systems would be 6 months minimum. Because of time and budgetary constraints, this proposed design solution was rejected.

Alternatives Investigated

A second potential solution under consideration involved acquiring a National Instruments Slot 0 PC that would be installed directly in the testing instrument rack. The result would be an aesthetically pleasing integration of the testing system. However, time and money were again major considerations in elimination of this proposed solution. The Slot 0 PC costs would exceed $5,000 and requires a 3-month lead time for procurement. Since the design solution required a great deal of program writing and debugging, the 3-month lead time would not allow for project completion in the time allotted.

Figure 6.5. Example of a technical report

The third and final alternative the design team considered was a system based solely on existing hardware and software. This option required use of a Windows NT PC equipped with several digital input controller PCI cards capable of handling output from National Instruments signal conditioners. This option involved writing software interfaces between the data collected through National Instruments LabVIEW programming language and a C++ program written by the design team. A second software interface was required between the output from the C++ program and an Excel spreadsheet. Although this design solution was less elegant than the other two possibilities that were considered, cost and time considerations were optimal, so this option was selected for implementation.

Selection of
Optimal Solution

Data Acquisition System

A sport-utility vehicle (SUV) was used for the completion of this project. The SUV had all the necessary sensors installed on it, and the rear seats were removed to accommodate the testing equipment. The sensors on the vehicle were for cylinder pressure (8), cylinder temperature (8), angular rotation of the crankshaft (SODEP), exhaust manifold temperature (left and right), catalytic converter temperature, and muffler temperature. Figure 1 shows the locations of the sensors on the vehicle and Figure 2 shows the logic diagram for the data acquisition system. The output from each sensor was converted from analog to digital through a signal conditioner and was input into various channels on the data acquisition board mounted in the rear of the vehicle.

Description of
Operating
Environment

Figures 1 and 2
Referenced

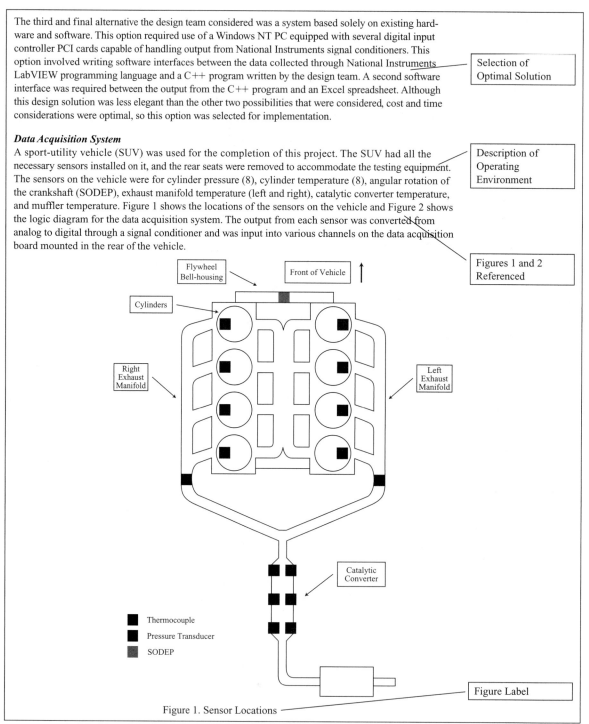

Figure 1. Sensor Locations

Figure Label

Figure 6.5. Example of a technical report (*continued*)

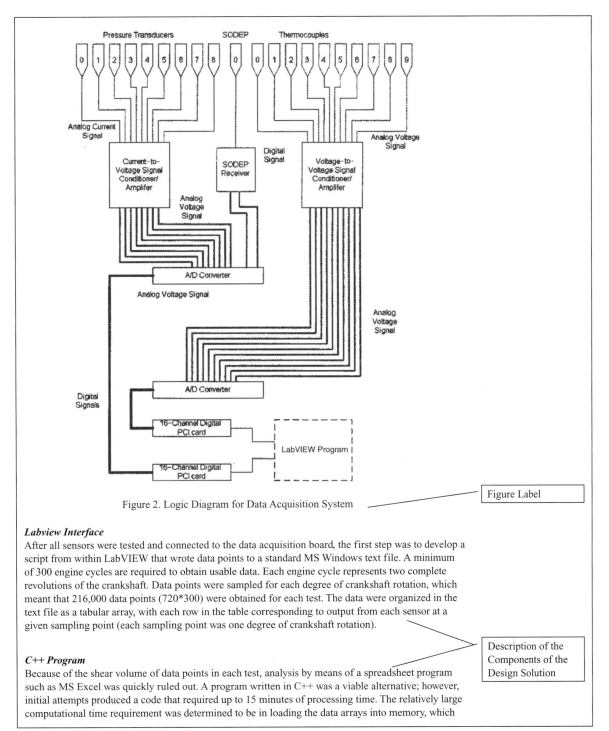

Figure 2. Logic Diagram for Data Acquisition System

Labview Interface
After all sensors were tested and connected to the data acquisition board, the first step was to develop a script from within LabVIEW that wrote data points to a standard MS Windows text file. A minimum of 300 engine cycles are required to obtain usable data. Each engine cycle represents two complete revolutions of the crankshaft. Data points were sampled for each degree of crankshaft rotation, which meant that 216,000 data points (720*300) were obtained for each test. The data were organized in the text file as a tabular array, with each row in the table corresponding to output from each sensor at a given sampling point (each sampling point was one degree of crankshaft rotation).

C++ Program
Because of the shear volume of data points in each test, analysis by means of a spreadsheet program such as MS Excel was quickly ruled out. A program written in C++ was a viable alternative; however, initial attempts produced a code that required up to 15 minutes of processing time. The relatively large computational time requirement was determined to be in loading the data arrays into memory, which

Figure 6.5. Example of a technical report (*continued*)

accounted for 90% of the total run time. The use of dynamic memory allocation with pointers reduced the computational time to a mere 45 seconds.

Pressure data from each cylinder were used to calculate the gross mean effective pressure (GMEP), the pumping mean effective pressure (PMEP), and the nominal mean effective pressure (NMEP) by using standard numerical formulations [1]. Cylinder pressure data were also used to calculate the burn rate for each [1]. (The *burn rate* is defined as the time required for 50% of the air-fuel mixture to be consumed.) The output from this analysis was input into an Excel spreadsheet through the use of an embedded Visual Basic script. Once the data are in the spreadsheet, various graphs can be created to display the data according to user needs.

Description of the Components of the Design Solution

Reference Cited

A graphical user interface (GUI) was also developed to make the analysis system more user friendly. The GUI allows the user to select the number of cylinders, the engine displacement, the polytropic expansion coefficient, the input file name, and the output file name.

The emissions are in the form of nitrous oxide (NO) and are a result of the combustion occurring inside the cylinders. The raw pressure and temperature, data obtained from the testing; the calculated values of GMEP, PMEP, NMEP, and burn rate: and the Ideal Gas Law were used in the standard-state Gibbs function to yield the following [2]:

Equation label

$$\frac{dX_{NO}}{dt} = \frac{RT}{P}\frac{d[NO]}{dt}\,(\text{ppm/s}) \tag{1}$$

where X_{NO} is a term relating the mass of NO to its molecular weight, R is the universal gas constant, T is the temperature in degrees Kelvin, P is the pressure, and NO is the amount of NO emitted during combustion. A routine was written within the C++ program that numerically integrated Equation 1 and output the results to the Excel spreadsheet for graphing. Figure 3 shows sample output in the form obtained through this analysis.

Variables Defined

Results Displayed

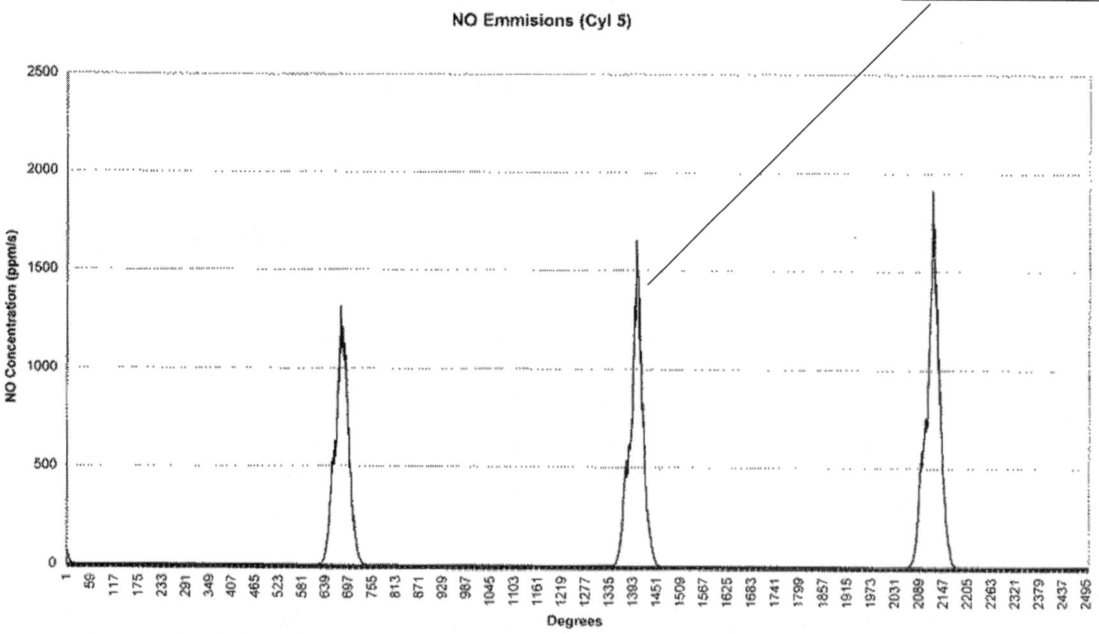

Figure 3. Sample Output from Data Analysis System

Figure 6.5. Example of a technical report (*continued*)

Conclusions

Through this project, a data analysis system for estimating NO emissions from vehicle cylinders was successfully developed. This system enables accurate prediction of emissions without the expense of a dynamometer chamber. The system uses a simple Windows NT PC platform for analysis and standard sensors and software for data acquisition. The system is cost effective and user friendly, and it should greatly reduce the time and expense involved in the design of new vehicle models.

Recommendations

Future enhancements of this data analysis system would be to improve the user interface for performing the analysis and to streamline the system into a single routine. Another enhancement would be to migrate the system to the MTS/CAS system so that additional data collected throughout the vehicle could be incorporated into the analysis.

> Future Enhancements to Project Cited

References

1. SUV Motor Company. 2000 Jan. 12. Acquisition and analysis of cylinder pressure data. Internal publication. Chicago: SUV Motor Company.
2. Ramos JI. 1989. Internal combustion engine modeling. New York: Hemisphere. 422 p.

> References Are Often on a Separate Page.

Figure 6.5. **Example of a technical report (*continued*)**

Form. Usually, you divide experimental reports into a number of sections, each of which is preceded by a heading, similar to the way you create a technical design report. A description of the sections you might include in a typical experimental report follows.

Title and author. Frequently, the title and author (or authors) are included on a separate title page, but a separate page is not always necessary. The title should help people decide whether they want to continue reading the report. The date of the report should also be included.

Abstract or summary. For industry reports, an abstract or an executive summary is generally required.

Introduction or purpose. The Introduction or Purpose section contains a short paragraph in which the purpose of the experiment is stated. In some instances, a summary of the experimental results may be provided in this section to let the reader know upfront what the major findings were.

Procedure. The Procedure section describes the experimental procedure that was followed. Typically, you use the passive voice and do not use first person in this section of the report. If the procedure you followed is in accordance with published standards, cite the appropriate standard in this section. However, if the procedure you followed deviated from the published standards, cite any differences instead.

Basic theory. In certain types of reports, the Basic Theory section may be optional. If it is included, theories from books or journal articles, as well as governing equations describing the theory, should be discussed in this section.

Data and results. If you have a lot of experimental data, tabulate it logically so that readers can understand it quickly and easily. Sometimes, including a graph to show trends in the data is also useful. (Tables and figures are covered in Chapter 9.) Large amounts of data are often put in an appendix rather than in the body of the report.

Discussion. In the Discussion section, you describe the data and results presented in the previous section of the report. If reference values are available for any of the experimentally determined quantities, they should be presented and properly cited in this section. You should also compare experimentally determined values with the reference values. In this section, you should mention any sources of experimental error in the data and any problems you had during the experiment. If the data are unusual in any way, you should discuss the unusual aspects of the data and state possible reasons why your data are different than expected. Sometimes, combining the previous section and this one into a single section titled "Results and Discussion" is acceptable. This practice is particularly appropriate when the data being presented are complex and require discussion as they are presented to make them understandable to the reader.

Conclusions. In the Conclusions section, you state any conclusions that can be drawn from the experimental data. Further areas for investigation should also be presented as appropriate. No new material should be presented.

Recommendations. If the conclusions from the experimental study require any specific actions, these actions should be stated in the Recommendations section. This section can be a single sentence to a page or two long. Often, the specific recommendations are listed.

References. You should include a list of references if you cited any in your report.

Appendixes. Include appendixes with your report as needed to display data or results.

Mechanics. As with technical design reports, headings and subheadings should be distinguished from the remainder of the text by changes in font size or treatment. In addition, several key aspects of experimental reports that are specific to this type of documentation should be included in them:

■ ***Compare your results with reference values.*** Many of the properties and results you obtain experimentally will have corresponding values in books or articles. For example, if you experimentally determine the density of a given material, you should look up the density in a reference book for comparison. When you compare your results with reference values, you should state the source of such values and the percentage of deviation between your results and the reference values (citations are covered in Chapter 3).

■ ***Explain any results that do not correspond well with the reference values.*** If your results differ from the reference values by more than a few percentage points, you should discuss possible reasons for this difference. Do not list "experimental error" as the only reason for the deviation. Try to list some of the sources of the experimental error or other possible reasons for the deviation.

■ ***Use tables or graphs to display large amounts of data.*** Try not to bury the results within the text of the report. Using tables or graphs can make your data easy to locate and easy to grasp.

■ ***Include sample calculations to illustrate key points.*** Sometimes, calculations are best included as an appendix to the report; however, when the calculations are not standard, you may want to include them in the main body of the report.

Example
Figure 6.6 shows a sample experimental report.

6.3.3 Progress Reports

Function. As an engineering student, you will often be assigned semester-long design projects. These projects are partially meant to simulate the working world, where most projects you work on as an engineer will require several months or even years before they are completed. *Progress reports* are used to inform your boss, your client, or your professor about your progress on a given project to date and to alert her about potential problems that may cause delays in its successful, timely completion.

Form. A good progress report gives specifics about exactly what has been completed on the project, how it was accomplished, potential problems that may arise in the next stages of the project and how they could or should be addressed, and any deviations in project progress from the original time estimates (i.e., whether the project is on time, ahead of schedule, or behind schedule). A Gantt chart (described in Chapter 9) is often included with progress reports so that your boss or client can see graphically how the project is progressing toward completion.

Since your boss or professor typically knows what your project entails, you do not usually need to describe the project in detail to him in your progress report. However, most large projects can be divided into small, manageable "chunks," and a good progress report will state the progress made toward meeting

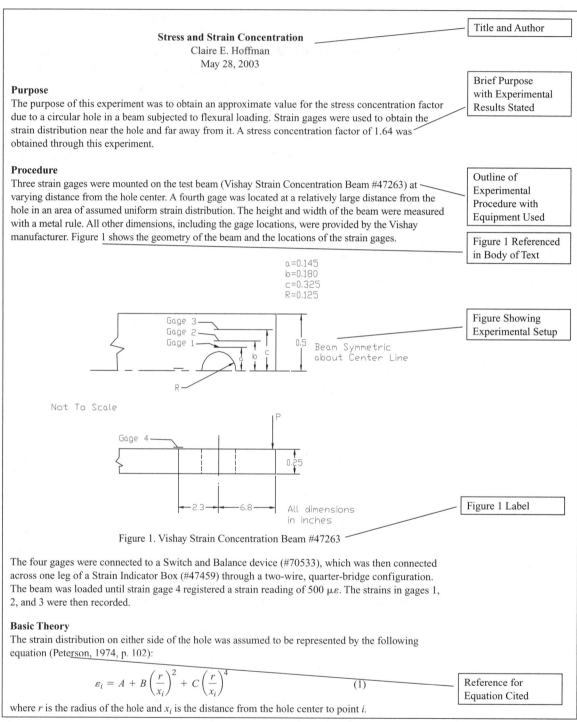

Stress and Strain Concentration
Claire E. Hoffman
May 28, 2003

Title and Author

Purpose

The purpose of this experiment was to obtain an approximate value for the stress concentration factor due to a circular hole in a beam subjected to flexural loading. Strain gages were used to obtain the strain distribution near the hole and far away from it. A stress concentration factor of 1.64 was obtained through this experiment.

Brief Purpose with Experimental Results Stated

Procedure

Three strain gages were mounted on the test beam (Vishay Strain Concentration Beam #47263) at varying distance from the hole center. A fourth gage was located at a relatively large distance from the hole in an area of assumed uniform strain distribution. The height and width of the beam were measured with a metal rule. All other dimensions, including the gage locations, were provided by the Vishay manufacturer. Figure 1 shows the geometry of the beam and the locations of the strain gages.

Outline of Experimental Procedure with Equipment Used

Figure 1 Referenced in Body of Text

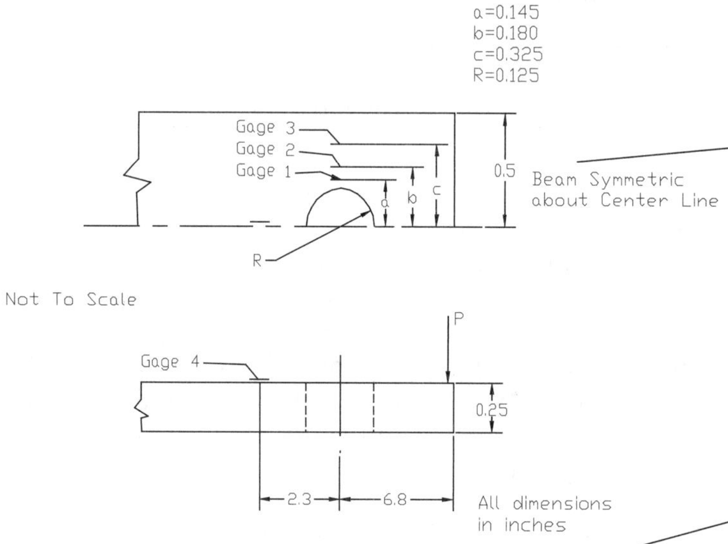
a=0.145
b=0.180
c=0.325
R=0.125
Gage 3
Gage 2
Gage 1
a
b
c
0.5
Beam Symmetric about Center Line
R
Not To Scale
P
Gage 4
0.25
2.3
6.8
All dimensions in inches

Figure Showing Experimental Setup

Figure 1. Vishay Strain Concentration Beam #47263

Figure 1 Label

The four gages were connected to a Switch and Balance device (#70533), which was then connected across one leg of a Strain Indicator Box (#47459) through a two-wire, quarter-bridge configuration. The beam was loaded until strain gage 4 registered a strain reading of 500 $\mu\varepsilon$. The strains in gages 1, 2, and 3 were then recorded.

Basic Theory

The strain distribution on either side of the hole was assumed to be represented by the following equation (Peterson, 1974, p. 102):

$$\varepsilon_i = A + B\left(\frac{r}{x_i}\right)^2 + C\left(\frac{r}{x_i}\right)^4 \qquad (1)$$

Reference for Equation Cited

where r is the radius of the hole and x_i is the distance from the hole center to point i.

Figure 6.6. Sample experimental report

In addition, the strain at gage location 4 was assumed to represent the nominal stress of the reduced section. For this assumption to hold true, the following equation must be satisfied:

$$L_h/L_4 = (b - d)/b \qquad (2)$$

where L_h and L_4 are the distances from the load to the hole center and gage 4, respectively; b is the beam width; and d is the diameter of the hole. In this case, both sides of the equation are equal to 0.75, so this assumption is satisfied. Other assumptions used in this experiment were that gage 4 was located in a uniform stress field and that the elastic limit of the beam was not exceeded.

Data and Results

The data obtained in this experiment are presented in Table 1.

Table 1. Experimental Data

Gage 1	Gage 2	Gage 3	Gage 4
638 $\mu\varepsilon$	520 $\mu\varepsilon$	466 $\mu\varepsilon$	500 $\mu\varepsilon$

Discussion

Strain readings and appropriate dimensions for gages 1–3 were substituted into equation (1), which resulted in three independent equations to solve for the constants A, B, and C. For this experimental procedure, the following were determined: A = 477.0 $\mu\varepsilon$, B = −146.5 $\mu\varepsilon$, and C = 488.6 $\mu\varepsilon$. On the basis of the form of the equation, the maximum strain will be obtained when $x = r$ (i.e., at the "edge" of the hole), and the minimum strain will occur at $x \cong \infty$. From this experiment, $\varepsilon_{max} = 819$ $\mu\varepsilon$ and $\varepsilon_{min} = 477$ $\mu\varepsilon$ were obtained.

According to Peterson (1974), the stress concentration factor for this experimental setup should be 1.84. From the calculation of ε_{max}, a stress concentration factor of 1.64 was obtained (i.e., 819 $\mu\varepsilon$/500 $\mu\varepsilon$) which represents a value 11% less than the reference value.

Several sources of error were possible in this experiment. One was that only one reading was obtained; therefore, any errors due to faulty connections or other problems were likely not detected. A second possible source of error was that in the area of assumed uniform strain distribution, only one gage was used, which meant there was no way of determining whether this assumption was valid. A third possible source of error was that a two-wire quarter-bridge hookup was used, which meant that any thermal strains would not be accounted for in the gage readings. The thermal strains were assumed to be negligible.

Conclusions

Through this experiment, a stress concentration factor of 1.64 was obtained for the beam in question. This value is 11% less than the reference value for the stress concentration factor. Since the purpose of the experiment was to determine an *approximate* value for the stress concentration factor for the beam, a value of 1.64 seems reasonable.

References

Peterson, R. E. (1974). *Stress concentration factors*, New York: Wiley.

Figure 6.6. Sample experimental report (*continued*)

each smaller goal. Progress reports are typically written at designated times throughout the course of the project. Sometimes, weekly or biweekly progress reports are required. Other times, quarterly reports are sufficient. Filing your progress report on time will greatly assist your boss, who may be managing several engineering teams and several projects at once.

Mechanics. Memos are the most common type of document used to file a progress report; however, some companies may have standard forms required for all projects. If you are filing a progress report with a client instead of a boss, a letter is likely the best format to select. Since a progress report details the progress you (or your team) have made on a given project, use of first person (*I, we, me, us*) is generally acceptable. If your progress report is a memo, make sure you include the date, for documentation purposes, and in the memo subject line (RE:), insert something such as this:

RE: Progress Report—Bangor Sewage Treatment Facility Project

6.3.4 Technical Papers

Function. The function of a *technical paper* is similar to that of a technical report; however, the audience is usually broader than just your particular company. Most of the time, technical papers are written for journals or conferences. The purpose of a paper is to report the results of a technical study you undertook. Some papers will describe the outcome of experimental work, while others will focus on results from theoretical work. Another type of technical paper is known as a *survey paper,* or *state-of-the-art paper.* It comprehensively lists and discusses the results of a range of previously published research studies in a given field over an extended period.

Form. The form of a technical paper is similar to that of a technical report. Once again, your paper should be divided into several sections or subsections with appropriate headings. Although the format of a technical paper will vary depending on context and audience, a general format for a paper is as follows.

Abstract. An abstract summarizes the contents of your paper and includes basic background information, major findings, and conclusions. Abstracts are discussed more fully elsewhere in this chapter.

Introduction or background. Depending on the nature of your specific study, you should include a section that introduces the subject and describes relevant background information. This section should help the reader learn about theories or previous studies related to what you have done. In this section, you will generally need to cite the work of previous researchers or engineers. (Citations are covered in Chapter 3.) In many technical papers, the Introduction and Background sections are separate. In this case, the Introduction briefly (no more than two double-spaced, typewritten pages) describes what the study was intended to do, why it was important to do it, and how it was done. Often the last paragraph of the Introduction is a statement of the objectives of the study. The Background section consists of a review of relevant background information with proper citation of past work. Most likely, you will need to go to the library or look through company archives to search for the information to be included in this section of the report.

Current study. In the Current Study section, you describe the methods you used. If your paper is written to highlight results from an experimental study, include an outline of the procedure you followed. If your paper highlights theoretical work you have accomplished, describe the formulation of your theories. In either case, the contents of this section will be an abbreviated form of a report you may have written on this study. For example, if you are writing a paper describing an experimental study, the procedure included in your paper will be described in much less detail than if you were writing a lab report on the experiments. Sometimes fairly detailed theoretical derivations are included in an appendix. The Current Study section may also consist of more than one main heading. For instance, in the case of a study that included both an experimental portion and a theoretical portion, the sections might be titled "Experimental Procedure" and "Theoretical Development," respectively.

Results and discussion. The Results and Discussion section combines results along with a discussion of the results. Readers are much more interested in what your results mean than just the results. You do not include every single data point you may have gathered—only major findings and results. Sometimes, including the data in an appendix is appropriate. Figures, charts, tables, and graphs are typically used throughout this section to illustrate your results and analysis. These forms of graphical communication are covered in Chapter 9.

Conclusions. In the Conclusions section of your paper, restate the major findings of your study clearly and concisely. You will have to reduce the several paragraphs in the Results and Discussion section to one or two sentences in this section. No new material should be presented in this section.

Recommendations. Often the work described in a technical paper suggests some logical future efforts that could extend the knowledge base beyond that resulting from the current study. In this section, you give recommendations for a few possible studies you or another investigator might undertake. The recommendations are sometimes numbered. This section may also be titled "Recommendations for Future Work" or simply "Future Work."

References. Technical papers should include reference to the previous work of other researchers. These references are usually cited in the Background section of the paper; however, some may be cited in the Results and Discussion section also. If you use an Introduction section separate from the Background, you may also cite references in the Introduction section, particularly when discussing the importance of the project. (Citing references is covered in Chapter 3.)

Appendixes. If appendixes are allowed by the publisher, include them in your paper as needed to display data or results, but only if they are necessary. For papers, limit the amount of appendix material as much as possible.

Each main section of a paper typically has one or two subsections, depending on the topic and the nature of the study you undertook.

Mechanics. All the rules for making headings and subheadings of a report stand out from the rest of the text also apply to technical papers. You typically use a relatively formal writing style for a paper. For example, when writing a technical paper, you should avoid first and second person (*I* or *you*), use the passive voice, and make sure you follow all the formal grammatical rules (e.g., try to avoid using a preposition at the end of a sentence).

Example

Figure 6.7 shows an example of a technical paper.

MODELING WOOD-PEG-CONNECTED TIMBER FRAMES

William M. Bulleit

Abstract. *Wood-peg-connected timber frames require the use of an analysis procedure that accounts for the connection behavior. The peg stiffness can be modeled by using short elements whose axial stiffness is determined by assuming that the peg acts as a simply supported beam with a concentrated load at midspan. The concentrated load is the tenon bearing on the peg and the supports are the sides of the mortise. This approach has been shown to work when its data have been compared with test data on the behavior of a single connection. This modeling approach has been extended to a fairly large frame to examine the sensitivity of the frame to various modeling assumptions, particularly the effects of member-to-member contact at a joint. If the effects of contact are not accounted for, estimations of some load effects may be underestimated by more than 100 percent, although most underestimations are 25 percent or less.*

Keywords: Mortise and tenon, timber engineering, heavy timber, wood engineering.

1. INTRODUCTION

The need to understand the structural behavior of wood-pegged timber connections has become more pressing during the last decade or so. This need has been driven by the trend toward renovation and rehabilitation of historic wood structures and the significant increase in the use of traditional methods in the construction of new structures.

The literature on the behavior and analysis of traditional timber frames is limited. Benson and Gruber [1] discussed structural analysis and design in an elementary fashion and only for members. The analysis of frames was not addressed. Brungraber [2] tested joints and a few frames, performed 2–D finite element analyses on some joint details, and proposed a three–spring joint model for frame analysis. He compared his model with his frame tests with good results. Later work by Weaver [3] showed that Brungraber's frames were insensitive to joint behavior, so almost any reasonable model worked well for predicting their behavior. Kessel et al. [4] reported on the reconstruction of an eight-story timber frame building in Germany. They modeled the joints as pinned connections, but no tests were performed to examine the adequacy of the model. More recently, Kessel and Augustin [5, 6] published their experimental studies on the load capacity of wood-pegged joints.

The objective of this paper is to examine the analysis of timber frames that are more complicated than those examined in the earlier work.

2. ANALYSIS TECHNIQUE

2.1 Analysis Guidelines

On the basis of past timber frame research, some guidelines for frame analysis have been developed [7,8,9]:

Callout boxes:
- Overview of paper contents with results stated up front
- Abstract font treatment sets it apart from the rest of the document
- Keywords supplied for indexing
- Numbered referencing system
- Overview of what others have done

Figure 6.7. Example of a technical paper

- Frame members should be modeled as beam columns with the effects of shear included.
- Joints should be assumed to carry no moment (i.e., they are free to rotate).
- Eccentricity of force should be included. For instance, when the pegs are not located at the centroid of a column (post) or beam, the eccentricity between the pegs and the column or beam centroid should be included in the analysis.
- Effects of contact should be considered. Where contact between one member and another could have a significant impact on the behavior of the frame, account for the contact.

> List showing parallel structure

The last guideline generally requires that the frame be analyzed more than once for each load case. First, an analysis must be performed to discern which members are in tension. Members in tension should have the axial stiffness of the connections on that member based on the peg behavior only (i.e., no contact effects), which is particularly important for members that act as braces (e.g., knee braces and collar ties). Second, braces in compression should be included in the analysis by accounting for the possibility that they may or may not be in contact with the adjacent member.

2.2 Analysis Approach

Consider the guidelines in section 2.1 as they pertain to analysis of the frame shown in figure 1. This frame is taken from Ref. [9]. To show how a joint is modeled, consider the post-beam connection circled and labeled *A* in fig. 1. The model of this joint is shown in figure 2a. The elements framing into the joint from the top and bottom, post elements 5 and 2, and the beam element, no. 8, should be modeled as beam-column elements. These beam-column elements should have the post or beam section properties, as appropriate, and should include the effects of shear in the stiffness matrix. The elements labeled 3 and 4 are beam-column elements that model the reduced cross section at the mortise. Some recent work [9] indicates that continuing the full post section properties into this region may be adequate. Element 6 accounts for eccentricity of force and is a very stiff element, on the order of 10,000 times stiffer than the mortise element (i.e., element 3 or 4), framing into it. The last element, no. 7, with nodes 19 and 20, models the section of the beam tenon from the face of the post to the pegs. At the pegs, node 19, there should be a moment release on this element that allows that end of the element to act as a hinge preventing the pegs from carrying a moment. The axial stiffness of this element is controlled by the state of stress in element 8. If element 8 is in compression, contact may be assumed and the axial stiffness becomes the same as that of the beam (i.e. element 8). If the beam is in tension, then the axial stiffness of element 7 is controlled by the pegs.

> Description of the analysis and discussion of the results obtained

The pegs are modeled as short beams with a concentrated load acting at the center of the tenon. The span of the short beam is assumed to be the distance between the middle of the mortise walls on each side of the post. The axial stiffness, k_a, of element 7 would then be calculated with this equation:

$$k_a = 3(48EI/L^3) \tag{1}$$

where 3 is the number of pegs, E is the modulus of elasticity of the peg, I is the moment of inertia of the peg, and L is the peg span described previously. The other properties of this element should be based on the tenon dimensions. Again, there are indications [7] that the properties, other than axial stiffness, may be based on the beam dimensions (i.e., a reduction in only the axial stiffness when the beam is in tension).

Next consider the knee brace/beam connection circled on figure 1 as *B*. This joint is shown in figure 2b. The two beam elements, numbers 8 and 12, and the short element, number 11, have their section properties determined in a manner similar to that described for the post. The short element with nodes 12 and 13 is used to model the peg effects. There is a moment release at node 12. The axial stiffness is based on the cross-sectional area of the knee brace when the knee brace is in compression and is in contact with the beam and post. It is based on the single peg in bending, as described previously, when the knee brace is in tension or if the knee brace is in compression and not in contact with the beam or post. The other properties are based on the knee brace cross section, although since there is a member release on each end of the knee brace, its behavior is controlled by the axial stiffness of the elements.

Figure 6.7. Example of a technical paper (*continued*)

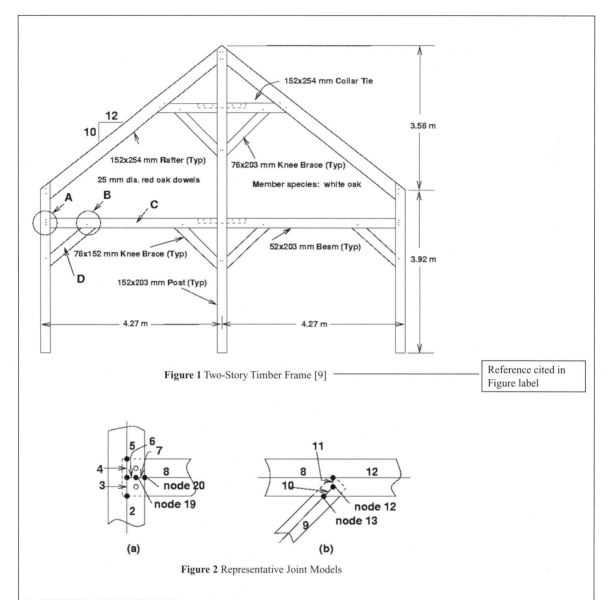

Figure 1 Two-Story Timber Frame [9]

Reference cited in Figure label

Figure 2 Representative Joint Models

3. RESULTS AND DISCUSSION

Considering design wind load, as if the wind were blowing *from the left* of the frame, with no floor or snow load, the tensile force in the left exterior knee brace, D in figure 1, is 340 N if contact is assumed for all members in compression and is 420 N if knee braces in compression are assumed not to be in contact. This difference is potentially significant, particularly since it affects the pegs connecting the knee brace to the post and beam.

The load effects that are sensitive to contact are also sensitive to peg stiffness. The simple beam model described in this paper for modeling peg stiffness worked well when it was used in analysis of

Figure 6.7. Example of a technical paper (*continued*)

test frames [8], but recent single-peg joint tests [10] showed that the effective span is longer than that predicted in the approach just described. This means the model used in this study might overestimate the stiffness of the pegs. The effect of this overestimation was relatively small (<5%) for load effects examined in the frame discussed here.

4. CONCLUSIONS

A technique for analyzing traditional timber frames by using commercial structural analysis software has been developed. Initial studies indicate that multiple analyses are required for each load case to bracket load effects.

> Conclusions stated concisely and without introducing any additional information

5. REFERENCES

> References listed in the order in which they appear in the paper

[1]   Benson, T.; Gruber, J., *Building the timber frame house*, Charles Scribner's Sons, New York, NY, 1980.

[2]   Brungraber, R. L., *Traditional timber joinery: A modern analysis*. Ph.D. dissertation, Stanford University, 1985.

[3]   Weaver, D. A., *Modeling the behavior of traditionally connected timber frames*, Masters thesis, Michigan Technological University, 1993.

[4]   Kessel, M. H.; Speich, M.; Hinkes, F.–J., "The reconstruction of an eight–floor timber frame house at Hildesheim (FRG)", *Proceedings of the 1988 International Timber Engineering Conference*, Vol. 1, 1988.

[5]   Kessel, M. H.; Augustin, R., "Load behavior of connections with pegs I," Trans. by Peavy, M. and Schmidt, R., *Timber Framing,* No. 38, 1995.

[6]   Kessel, M. H.; Augustin, R., "Load behavior of connections with pegs II," Trans. by Peavy, M. and Schmidt, R. *Timber Framing,* No. 39, 1995.

[7]   Sandberg, L. B.; Bulleit, W. M.; O'Bryant, T. L.; Postlewaite, J. J.; Schaffer, J. J., "Experimental investigation of traditional timber connections", *Proceedings of the 1996 International Timber Engineering Conference*, Vol. 4, 1996.

[8]   Bulleit, W. M.; Sandberg, L. B.; O'Bryant, T. L.; Weaver, D. A.; Pattison, W. E., "Analysis of frames with traditional timber connections", *Proceedings of the 1996 International Timber Engineering Conference*, Vol. 4, 1996.

[9]   Drewek, M. W., *Modeling the behavior of traditional timber frames,* Masters thesis, Michigan Technological University, 1997.

[10] Reid, E. H., *Behavior of wood pegs in traditional timber frame connections,* Masters thesis, Michigan Technological University, 1997.

Figure 6.7. Example of a technical paper (*continued*)

6.4   Specifying Design and Assembly Instructions

In the final stage of the design process—design implementation—engineers build, manufacture, or implement the final product. Typically, a different group of engineers carries out the implementation phase than that which carries out the design phase. At a large company, different divisions are generally in charge of manufacturing the necessary parts and assembling the final product. At smaller firms, machining companies may be contracted to create the individual parts, and the assembly may be done at the original firm. In building and road construction projects, the design engineering firm usually hires a general contractor to implement the design. The general contractor in turn hires subcontractors who are each responsible for a portion of the entire project. Because engineers who were not involved in the design are often responsible for the project implementation, clear communication among these parties is critical.

6.4.1 Design Specifications

Function. *Design specifications* are usually written for large-scale construction projects. Typically, these projects are one-of-a-kind designs constructed by a general contractor who oversees several subcontractors hired to perform specific tasks on the project. The documentation for a large-scale construction project consists of at least two parts:

1. A set of drawings that graphically illustrate the design solution
2. A set of written specifications that supply any information not provided by the drawings

The specifications accompany the drawings, and these two items make up the entire design documentation. Because the specifications are *written* documents, they take precedence in any legal proceedings that occur as a result of the building contract. A third common type of documentation is a written contract between the engineers who carried out the design and the engineers responsible for the project implementation.

For example, suppose you are a chemical engineer responsible for the design and construction of an oil refinery. During the design process, several engineering teams design all the systems for the entire plant. A civil engineer designs the plant structure, a mechanical engineer designs the pumping and the heating and cooling systems, an electrical engineer identifies the power needs of the plant, and an environmental engineer develops the required pollution-prevention mechanisms. As the chemical engineer, you are responsible for determining how the material flows through the plant, the temperature that should be maintained at each stage of the chemical processing, and the different compounds to be extracted or added to the product during processing.

After the plant is designed, a general contractor is hired to build the facility. The contractor in turn hires subcontractors to construct the various systems within the plant. During all this construction, explicit written communication is supplied to the contractor and subcontractors in the form of design specifications. Specifications include provisions for items such as the following:

- The size and capacity of all pumps used in the plant
- The material characteristics and forming techniques for the concrete and other materials in the plant
- The government codes, standards, and regulations that must be followed
- The terms and responsibilities for hiring subcontractors
- The soil compaction required for the foundation of buildings within the plant

- The type of cabling to be used in the wiring systems
- The required overall performance characteristics of the various plant systems

Form. Two basic methods are used to write design specifications. Using one method, you specify the techniques and materials to be used during the construction process. Using the other method, you specify the desired result and leave decisions about methods and materials to the contractor or manufacturer, as long as the performance requirements are achieved. In either case, you should use clear language and write in a simple style so that the contractor, the subcontractors, and the job supervisors understand exactly what you mean. Some large firms may have employees responsible for writing all company design specifications.

Design specifications should be written as if they were instructions to the contractor about what must be done. For example, you might say, "The job site shall be maintained in a neat and orderly condition and kept free from waste materials during the entire construction period." Design specifications presuppose that the contractor is performing the work. Therefore, the phrase *the contractor* is never the subject of the sentence; instead, the material or method is the subject. In the previously stated example, *the job site* is the subject of the sentence; the contractor is understood to be the party responsible for keeping the job site clean.

Mechanics. The number and type of specifications you write for a given design solution vary depending on the complexity of the project, its function, and the governing federal, state, or local regulations. The range of information required makes specifications an extremely important aspect of the design. In general, design specifications should include the following elements.

General requirements. The General Requirements section includes the nonlegal and nontechnical requirements for the job, such as the completion date for the work.

Reference sources. Materials used in construction projects must meet minimum standards developed by various governmental agencies. When you specify a material to be used for a construction project, you include the specific government regulation that applies to the material so that the contractors know where to look for this information when they purchase or test materials.

Material evaluation techniques. Specifications in the Material Evaluation Techniques section describe in detail the procedures to be followed in the evaluation of materials and the expected performance requirements.

Manufacturers or sources for materials. Sometimes you may want to specify a source for materials or products. For example, if you wanted to use a specific brand of pumps or other electromechanical components in the construction, you could include this information in the specifications.

Provisions for change. During construction, deviations from the original plans and specifications will occur. For example, as the oil refinery is constructed, some of the pipes may interfere with one another, which will require modification of the plant design. You may include provisions for tolerances and change as a part of the specifications.

Contractor's and subcontractors' limits. In the Contractor's and Subcontractors' Limits section, the specifications delineate the responsibilities of the various contractors and subcontractors on a given job.

Insurance requirements. Contractors and subcontractors are required to carry insurance covering worker's compensation, fire damage, liability, and so on. You outline the provisions for this insurance and the minimum insurance standards in the Insurance Requirements section of the design specifications.

Design criteria. Use the Design Criteria section to specify when a drawing cannot adequately show all the design criteria. For example, you can specify the type of door finish to use or the number of coats or thickness of paint for the walls in this section.

Example

Let's consider a set of construction project specifications. This example is based on the Construction Specifications Institute's (1996) *Manual of Practice*. The format described is widely accepted and its universalness aids communication between the specification writer and the specification user. The overall format consists of the following:

- Introductory information
- Bidding requirements
- Contracting requirements
- Facilities and spaces
- Systems and assemblies
- Construction products and activities

The Construction Products and Activities section is then broken down into 16 divisions:

1. General requirements
2. Site construction
3. Concrete
4. Masonry
5. Metals
6. Wood and plastics
7. Thermal and moisture protection
8. Doors and windows

9. Finishes
10. Specialties
11. Equipment
12. Furnishings
13. Special construction
14. Conveying systems
15. Mechanical
16. Electrical

Under each division are various sections. For instance, under division 7 are sections 07050, Basic: Thermal and Moisture Protection Materials and Methods, through section 07900, Joint Sealers. These major sections can be further subdivided. For example, under section 07900, Joint Sealers, is section 07920, Joint Sealants, which could include information on backer rods, caulking, joint fillers, and sealants. A single project may not require all these divisions or sections, but they are available if needed. Each section is then broken into three parts:

Part 1—General

Part 2—Products

Part 3—Execution

As an example, consider section 07600, Flashing and Sheet Metal. The specific information in that section could be as shown in Figure 6.8.

Clearly, writing design specifications is an endeavor that requires more knowledge and information than can be given to you in a text such as this. The *Manual of Practice* (Construction Specifications Institute 1996) or a textbook on contracts and specifications, such as Rosen (1999), will give you more in-depth coverage of this material.

6.4.2 Assembly Instructions

Function. Design specifications are typically used for building large constructed facilities; however, many engineering projects involve assembling numerous products for use by the general public (e.g., an automobile, a TV, or a hair dryer). The products are typically assembled at a manufacturing plant, either by humans or by robots. Many times, the individual components in the system are manufactured elsewhere and then brought together for final assembly at a common facility. After you design a system to meet an engineering need and all the parts have been manufactured, someone needs to assemble the parts into the completed system. For successful assembly, you need to supply a set of instructions that can be followed to achieve the desired result. You are probably familiar with the concept of *assembly instructions* if you ever purchased a bike or a piece of furniture with

<u>**SECTION 07600**</u>   <u>**FLASHING AND SHEET METAL**</u>

PART 1—GENERAL

<u>RELATED DOCUMENTS:</u>

Drawings and general provisions of contract, including General and Supplementary Conditions and Division 1 specification sections, apply to work of this section.

<u>DESCRIPTION OF WORK:</u>

The extent of each type of flashing and sheet metal work is indicated on the drawings and by provisions of this section.

The types of work specified in this section include the following:

> Metal counter flashing, and base flashing (if any)

> Built-in metal valleys, gutters, and scuppers

> Miscellaneous sheet metal accessories

Integral masonry flashings are specified as masonry work in sections of Division 4.

<u>JOB CONDITIONS:</u>

Coordinate work of this section with interfacing and adjoining work for proper sequencing of each installation. Ensure best possible weather resistance and durability of the work and protection of materials and finishes.

PART 2—PRODUCTS

<u>FLASHING AND SHEET METAL MATERIALS:</u>

Sheet Metal Flashing/Trim:

Copper: ASTM B 370, cold-rolled except where soft temper is required for forming; 16 oz (0.0216″ thick) except as otherwise indicated.

Miscellaneous Materials and Accessories:

Solder: For use with steel or copper, provide 50-50 tin/lead solder (ASTM B 32), with rosin flux.

Fasteners: Same metal as flashing/sheet metal, or other noncorrosive metal as recommended by sheet manufacturer. Match finish of exposed heads with materials being fastened.

Elastomeric Sealant: Generic type recommended by manufacturer of metal and fabricator of components being sealed; comply with FS TT-S-0027, TT-S-00230, or TT-S-001543.

Roofing Cement: ASTM D 2822, asphaltic.

Figure 6.8. Construction specification example

SECTION 07600   **FLASHING AND SHEET METAL**

PART 3—EXECUTION

INSTALLATION REQUIREMENTS:

General: Except as otherwise indicated, comply with manufacturer's installation instructions and recommendations, and with SMACNA *Architectural Sheet Metal Manual.* Anchor units of work securely in place by methods indicated, providing for thermal expansion of metal units; conceal fasteners where possible, and set units true to line and level as indicated. Install work with laps, joints, and seams, which will be permanently watertight and weatherproof.

Bed flanges of work in a thick coat of bituminous roofing cement where required for waterproof performance.

CLEANING AND PROTECTION:

Clean exposed metal surfaces, removing substances that might cause corrosion of metal or deterioration of finishes.

Protection: Installer shall advise Contractor of required procedures for surveillance and protection of flashings and sheet metal work during construction, to ensure that work will be without damage or deterioration, other than natural weathering, at time of substantial completion.

Figure 6.8. Construction specification example (*continued*)

the label "Some Assembly Required" on the box. You may or may not have been successful at assembling the system you purchased, depending in large part on the quality of the instructions and drawings the manufacturer provided.

Form. Assembly instructions should begin with what is known as a *bill of materials,* or *BOM.* A BOM is typically set up in tabular form and lists all the parts, as well as the quantity of each part, required for a given assembly. Most times, assembly instructions also include what is called an *assembly drawing,* which graphically shows how all parts fit together in the completed system. (Assembly drawings are discussed briefly in Chapter 9.) As an example, following is a set of assembly instructions for stapling together two sheets of paper:

Bill of Materials

Item	Quantity
Stapler	1
Staples	Several
Paper	2

1. Place the sheets of paper together in the desired orientation.
   Note: Ensure that the edges of both sheets are even with each other.
2. Insert the upper right-hand corner of the sheets into the stapler mechanism.
   Caution: Make sure your fingers are clear of the mechanism before beginning the next step.
3. Press down firmly on the upper arm of the stapler through its entire stroke.
4. Remove the paper from between the arms of the stapler mechanism.
   Result: Two pages stapled together in their upper right-hand corner.

Mechanics. Graphics are usually an important part of assembly instructions. Imagine how difficult assembling a vehicle out of building blocks would be without the pictures that accompany the instructions. Assembly instructions should be written as numbered lists, and only one task given in each list item. Each instruction should begin with a verb, as illustrated in the preceding example. Results, notes, and cautions should be included in the instructions as appropriate so that individuals completing the assembly know what to watch out for and whether they are achieving the desired intermediate results. However, you usually do not want to include these elements as part of the numbered set of instructions. Note that in the preceding example, they were included as separate items beneath the action items in each case. When you finish writing your assembly instructions, you may want to test them by giving them to an individual who is unfamiliar with the component you are working with and having her try to put it together. This suggestion does not apply for large industrial assemblages.

Example
Figure 6.9 shows an example of assembly instructions for an all-terrain vehicle made from building blocks.

6.5 Including Equations in Technical Documents

Function. Since mathematics is one of the languages of engineering, many forms of written technical communication commonly contain equations. Engineers use equations to describe physical phenomena and to solve design problems. As such, equations are often an essential component of effective technical communication.

Form. When including equations in your document, you must follow certain established rules. If your equation is relatively standard, you do not need to cite a

Assembly Instructions for an All-Terrain Vehicle Made from Building Blocks

Bill of materials with parts and quantities listed

Bill of Materials

Part	Quantity		Part	Quantity	
1 × 4 Brick	2	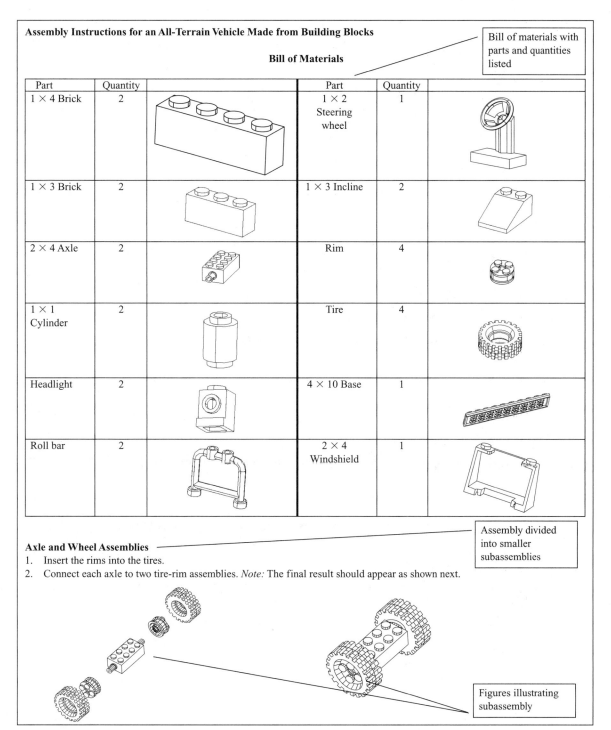	1 × 2 Steering wheel	1	
1 × 3 Brick	2		1 × 3 Incline	2	
2 × 4 Axle	2		Rim	4	
1 × 1 Cylinder	2		Tire	4	
Headlight	2		4 × 10 Base	1	
Roll bar	2		2 × 4 Windshield	1	

Axle and Wheel Assemblies

Assembly divided into smaller subassemblies

1. Insert the rims into the tires.
2. Connect each axle to two tire-rim assemblies. *Note:* The final result should appear as shown next.

Figures illustrating subassembly

Figure 6.9. Example of assembly instructions

Exhaust Assembly

3. Attach 1 × 1 Cylinders to Headlights to form "exhaust" pipes. *Note:* The final result should appear as shown next.

Body Assembly

4. Turn the 4 × 10 base over.
5. Attach the two axle assemblies to its bottom.
6. Connect the two exhaust pipes to the bottom of the base, at the rear end of the vehicle. *Note:* Your result should appear as shown next.

> Only one instruction given at a time

7. Turn the base-axle assembly over. *Note:* The assembly will now be "right side up."
8. Connect the 1 × 3 incline pieces across the front of the vehicle.
9. Attach one 1 × 3 brick and one 1 × 4 brick along each side of the base to form the vehicle sides. *Note:* Your result should appear as shown next.

> Results noted

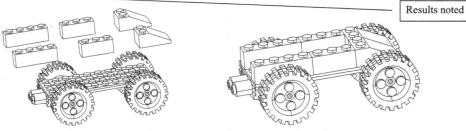

10. Attach the roll bars, windshield, and steering wheel to the base to complete your vehicle assembly. The final result should appear as shown next.

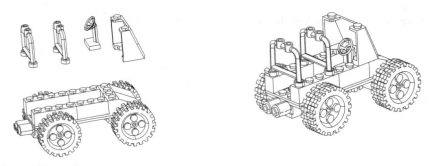

Figure 6.9. Example of assembly instructions (*continued*)

reference for its source; however, if the equation is unique, you must cite the source as described in Chapter 3.

When you include an equation, it should be centered on the page and numbered near the right margin. List all variables used in the equation in the text immediately following it. For example, your document could contain a paragraph similar to the following:

One well-established relationship is the stress-strain relationship for a linear-elastic member in uniaxial tension. This relationship is defined as follows (Riley et al., 1985):

$$\sigma = E\varepsilon \qquad\qquad \textbf{(Eq. 1)}$$

where σ is the stress, E is Young's modulus, and ε is the strain. The project described in this proposal will investigate the applicability of this relationship to some nontraditional materials.

The equation number in parentheses near the right margin sometimes appears as simply a parenthetical number; that is,

$$\sigma = E\varepsilon \qquad\qquad (1)$$

The format you should use depends on the instructions your company, the publisher of your document, or your professor gives you.

Mechanics. Fortunately, most modern-day word processors include an equation editor that allows equations to be easily inserted into your document. When inserting an equation, make sure you select standard symbols and variables so that the equation is easily understood by all who are familiar with the basic theory and mathematical models. After you include an equation and define the specific variables in it, you do not need to define them again if they appear in later equations in the document. In other words, each variable must be defined only once—at first mention. Note that writers may define some variables more than once if they believe doing so will make their document easier to read. For instance, if the first definition of the variable is many pages prior to the second equation in which it appears, redefining it at the second occurrence might be helpful to the reader.

Incorporating design calculations into a technical report is discussed in Chapter 7. Design calculations require a slightly different approach to including equations.

6.6 Exercises

1. Obtain a technical journal for your field and write an executive summary of one of the articles in it. Turn in to your instructor a copy of the article along with your summary.

2. Create checklists for the following types of documents:

   a. An executive summary
   b. A technical paper
   c. An experimental report
   d. A proposal
   e. Assembly instructions

3. Conduct an experiment in a building at your university (possibly the cafeteria or student union): For the people who come through a given entrance, count the number who turn right, turn left, go up or down stairs, and so forth. Write an experimental report that contains the data and the conclusions drawn.

4. Conduct an experiment in your home: Roll a single die or a pair of dice 50 times and record the frequency of each number. Write a report about this experiment. Compare your results with theoretically expected values.

5. Write a short proposal outlining a problem you see on your campus and how you would solve it.

6. A number of well-publicized engineering failures have occurred over the years. In your small group, choose one of the failures and research it in the library and on the Web. Write a paper about the failure, including its cause and the lessons learned.

7. Write a set of specifications about how to paint your room. Include a list that contains such information as the number of coats of paint, the application method, the color of paint, and the type of paint (i.e., enamel, latex, or semigloss).

8. Write a set of assembly instructions that someone could use to insert lead in your mechanical pencil.

9. Find a two- to three-page article in an engineering-oriented magazine (e.g., *Engineering News Record* or *Prism*), and write a 200- to 300-word abstract for the article.

10. After 2 to 3 weeks in a chosen course, write a progress report on what you have learned so far. The progress report should be in memo form and no more than one page long.

11. Using your word processor, reproduce the following text passage. (*Note:* You may have to consult the equation editor of your software to complete this assignment.)

> An electric circuit is forcing a time-varying current through an induction coil of 2×10^{-3} Henries. The voltage that will be developed is

$$v = L\frac{di}{dt} \qquad \text{(Eq. 1)}$$

where *v* is the voltage measured in volts, *L* is the inductance of the coil measured in Henries, *i* is the current measured in amps, and *t* is time in seconds. The power passed to and from the coil is

$$p = vi \qquad \text{(Eq. 2)}$$

where *p* is the power measured in watts. The energy passed to and from the coil is given by

$$w = \frac{1}{2}Li^2 \qquad \text{(Eq. 3)}$$

with energy, *w*, measured in joules.

6.7 References

Construction Specifications Institute. 1996. *Manual of practice.* Alexandria, VA: Construction Specifications Institute.

Library of Congress. 2004. *Subject headings.* 27th ed. Washington, DC: Library of Congress.

Rosen, H. J. 1999. *Construction specifications writing.* 4th ed. New York: Wiley.

7

Communication of Calculations

In most design offices, another engineer checks the engineering design calculations you perform. This check is a detailed examination of the design, including the basic assumptions, the design criteria, the correctness or appropriateness of the equations used, and the mathematics. Therefore, neatness and completeness are essential to your document so that someone who didn't do the work can understand your reasoning and follow your solution steps. Even if another engineer will not check your work, it will likely be used in either the body of the design report or one of its appendixes, which other individuals may review. Furthermore, the company usually keeps the design report and a set of complete calculations for future reference. Again, neatness and completeness are essential so that someone, maybe even you, can understand the calculations possibly years in the future.

The standard you should use in preparing design calculation documentation is this: *an engineer trained in your area of expertise should be able to understand your calculations without needing to ask you any questions or make any guesses about assumptions, design criteria, procedures, methods, or data that were used.*

7.1 Selecting Tools for Communicating Calculations

Design calculations are the one type of technical documentation for which electronic tools may not always be the best choice. Software tools are available that allow you to do engineering calculations so that the resulting printed documentation is in the format of traditional engineering calculations. However, for short problems or for problems that require several hand-drawn sketches, engineering calculations are still typically done with pencil and paper.

When creating this type of documentation by hand, you should use *engineer's paper,* which is almost universally accepted in the engineering profession as the paper of choice. Engineer's paper is usually light green with a heavy grid on its reverse side. You would normally never write on the reverse side of the paper: the grid exists merely to help you line up equations and sentences and to make sketching easier. Make sure your lettering is neat and your problem solution flows logically down the page. In the following sections, we outline the development of this type of technical communication in more detail.

Hand calculations on engineer's paper are probably suitable for most calculation documents you create for in-house use. If, however, you are preparing a report that will be distributed to a client or an outside agency, an electronic version of the design solution may be preferred. Several mathematical solvers on the market can do the job; however, be sure the final format of your document shows the solution to the problem in a step-by-step manner. Variables should be defined as appropriate, and figures used as needed to illustrate your problem solution fully. Alternatively, you can use a word processor and its built-in equation editor and drawing tool to create this type of documentation. This method is not as "elegant" as a mathematical solver, but it can be just as effective.

7.2 Communicating Calculations

Function. Design calculations support the design report and provide a detailed written record of the design process. A trained engineer should be able to obtain detailed information about the design from the calculations. The engineer could be working for your client or could be in your company. He could be looking at the calculations shortly after you perform them or years in the future. The engineer looking at the calculations could even be you in the future trying to determine what you did years in the past. In the worst case, he could be an engineer hired by the plaintiff's attorney in a lawsuit against your company or you personally. In this situation, missing details could be a serious problem. In *any* case of engineering failure that results in litigation, the court will likely try to attribute a percentage of the blame to each defendant so that costs and penalties can be allocated according to that breakdown. Omissions in your calculations can be construed to be sources of possible error that may cause you or your company to be considered partially or principally to blame for the failure.

Form. A range of formats can be used for documenting design calculations, but a key feature of all forms is that the calculations are presented in the order in which an experienced engineer would *perform* them. The design calculation documentation should include (a) all assumptions and design criteria, (b) references for physical property data, (c) design specifications, (d) design equations and models, (e) results from computer analyses, (f) sketches and notes, and (g) any supplemental information. As mentioned previously, your calculation documentation must be complete enough that another engineer can understand it without asking you any questions or needing to guess about any portion of the calculations.

Your company may have standardized sheets with a required format for calculations that are performed regularly. If so, you should use them. Even when no standardized sheets are provided, companies may require special calculation paper that guides you in creating your document. The type of information documented on this special paper typically includes such items as company name, designer's name, project, "object" being designed, account number, date, page number, and total number of pages. If your company does not use standardized sheets, an outline approach is often good for the overall format. Information such as company name, designer's name, and so forth should be placed on any calculation sheets you fill in, in standardized form or not.

The basic *form* for engineering calculations should be as follows:

- ■ *Information required to perform the analysis or design.* Such information includes a drawing to show the problem, material information, design criteria, design specifications, or any other necessary information.
- ■ *Problem solution organized to flow smoothly from top to bottom.* Any assumptions made must be clearly stated.
- ■ *Summary of the solution.* The summary could be a sketch or a highlighted summary statement.

Mechanics. The mechanics discussed in this section apply to the documentation of all design calculations in general, but one or more of the mechanics may be less critical in certain cases. For example, the general rule is that variables should be defined the first time they are used in your document. However, in disciplines in which specific variable names are used consistently or are specifically defined in a specification used in the design, definitions for these variables are generally not included. If you are in doubt about whether an item should be specifically stated in the documentation of design calculations, state it (i.e., err on the side of completeness). You can easily neglect to state something in your calculations that you already know, but remembering your original intent at a future date may be impossible if you neglect to explain items such as assumptions and approximations.

Description of design criteria and assumptions. If you are using design criteria or assumptions specific to the calculations, state them at the beginning of the documentation. If the criteria and assumptions are the same as stated in other sections of the design report, they need not be restated. If an assumption is specific to just a single portion of the calculations, stating that assumption at the point in the documentation at which it is used is acceptable. Design standards that will be adhered to in the calculations (e.g., American Society of Mechanical Engineers, or ASME, Pressure Vessel Code; American Concrete Institute, or ACI, Building Code) should be noted with the design criteria in the documentation.

Presentation of equations. Most calculations in the documentation will consist of equations and substitution of values into these equations. The equations and their use should be shown in an organized fashion throughout the document to enable another person to follow your work easily. An equation with variables should first be stated, followed by a repeat of the equation showing substitution of the numerical values of the variables in it. You should substitute the numerical values into the equation in the same manner in which they appear in the equation. For example,

$$x = \frac{1}{2} a t^2 + v_o t + x_o$$

$$x = \frac{1}{2} \cdot 4 \cdot 1.5^2 + 3 \cdot 1.5 + 5 = 14 \text{ m}$$

Note that in this equation, the units are placed with the answer but not with the individual terms in the equation. (Units are discussed subsequently.)

Another issue related to presenting the equations in your document occurs when equations exist within equations. Consider the following set of equations:

$$M = A_s f_y \left(d - \frac{a}{2} \right)$$

where

$$a = \frac{A_s f_y}{0.85 f_c' b}$$

In this case, one of the variables in the primary equation is defined by another equation. You should show both equations first, then repeat the second equation with values substituted to solve for *a,* and finally restate the first equation with the value for *a* and other numerical values to solve for *M:*

$$a = \frac{0.62 \cdot 60}{0.85 \cdot 4 \cdot 12} = 0.912 \text{ in.}$$

$$M = 0.62 \cdot 60 \left[21.81 - \frac{0.912}{2} \right] = 794 \text{ in.-kips}$$

Documentation of the appropriateness and origin of the equations. The calculations cannot stand by themselves; the source or sources for the equations you use should be documented. In other words, someone looking at your work should be able to determine where the equations were obtained from. If you stated the design standards you followed at the beginning of the calculations, undocumented equations will be assumed to have come from that document. The amount of detail required in equation documentation depends on the company you work for, the field you work in, or your instructor if you are in school. For instance, if you are designing steel structures and the firm always uses the American Institute of Steel Construction (AISC) Steel Design Code, you may not need to state this source in typical calculation documents. However, in a steel design class, your instructor may want you to show the section of the Steel Design Code from which you obtained a given equation.

One of the authors of this text (W.B.) worked for a firm where he designed small submersibles. A specific submersible could be designed to meet one or more of three sets of standards—American Bureau of Shipping, Det Norske Veritas, or Lloyd's Register of Shipping—with a full set of calculations performed to meet one of these for any submersible designed in the firm. Thus, the standard being followed for a given design needed to be stated at the top of the calculation documentation, and then all equations used in the document were assumed to come from that standard. If the submersible needed to be designed to a different set of standards, a separate set of calculations was required.

The difficulty of calculation documentation is increased if a design must be accomplished that adheres to more than one set of design standards. Suppose, for instance, you are designing a pressure vessel and its supporting structure. The pressure vessel may need to be designed to meet the ASME Pressure Vessel Code, and the supporting structure may need to be designed to meet the AISC Steel Design Code. In the documentation of this design, you must be sure to distinguish clearly between equations that originate from the ASME code and those that originate from the AISC code. In this example, another option might be to separate the pressure vessel design into a separate section of documentation relative to the support structure design. Then, you would simply reference the ASME code at the beginning of the pressure vessel design section and wait to reference the AISC code until the beginning of the support structure design section. Separation of your design solution like this may not always be possible; if not, the code citation must be shown each time a specific code is used.

If you need to derive your own equations for the design calculations, these derivations, *in their entirety*, must appear in the documentation. Clearly, many possible levels of equation documentation are possible. You need to follow your firm's or instructor's preferences first, but always remember the key criterion: *your calculation documentation must be complete enough that another engineer can follow your work without asking you any questions or needing to guess about any portion of the calculations.*

Definitions of variables and units. Units should be consistent throughout the documentation and should first appear when a variable is defined. The units should also be given with all results of equation solutions, as shown in the preceding examples. If you are following a set of design standards that specifies default units, your calculations should also use these units. This approach helps minimize errors associated with empirical and semi-empirical equations that may appear in the design standard. These types of equations may have constants associated with them that have implied units. For instance, the ultimate tensile stress of concrete can be determined by using the following equation:

$$f_t = 4.0\sqrt{f'_c}$$

where f_t is the ultimate tensile stress in psi and f'_c is the concrete cylinder ultimate compressive stress in psi. If the value of f'_c inserted into this equation does not have units of psi, the resulting answer will be incorrect. Note that the implied units of the constant, 4.0, are $psi^{1/2}$.

Variables should be defined the first time they are used in the calculation documentation, as they are in technical reports. A given variable name should also be used to define only one variable in a set of calculations. For example, if E is used as the variable name for the modulus of elasticity, it should not be used later in the documentation to represent energy. You could, however, use E_k for kinetic energy and E_p for potential energy, as well as E for modulus of elasticity, in a single document since the subscripts differentiate the variable names. If you are working within a set of design standards that uses specific variable names, you should use the same variable names in your documentation, and you need not define them in the calculations. Using variable names as specified in the standards is a fairly common practice in engineering offices; however, as usual, follow the standards set by your firm or your instructor when preparing your calculation documentation.

Physical property data. Physical property data obtained from design standards need not be referenced in the documentation so long as the design standard has been referenced. If the property data have been obtained from in-house testing, this fact must be stated in the calculation documentation, and the test results, or at least a summary of them, should be included or otherwise provided. Physical property data that have been taken from external sources must be documented and properly cited so that other individuals looking at your calculations know where to look to confirm your work. Citing external references in your calculation documentation is analogous to referencing sources in a technical report or paper, as described in Chapter 3. The referencing techniques described in Chapter 3 can be used for calculation documentation purposes also.

Figures and notes. Figures are often vital to explaining calculations and thus are an important part of the documentation. For instance, when you are defining object dimensions (either numerical values or dimension variables) that you may need to input into equations, using a figure is the easiest, clearest way to do so. If you have defined the width of a beam as b, including this variable on a sketch ensures that anyone looking at the calculations knows immediately that you mean the width and not the height dimensions when you use b in an equation.

In many design calculations you make and document, the final result is the design of some *thing* (e.g., a beam or a driveshaft or a circuit board). For well-defined items such as these, using a figure to summarize your design is a convenient way to show the final design solution. The figures can often be sketches, even if the design documentation is created in an electronic format, since hand sketches are still generally the fastest way to include figures in your document.

Notes explaining your calculations, assumptions, and design intent should be included liberally throughout the document. The design summary could be a note rather than a figure in some cases. For instance, if you are sizing a pump for use in a chemical processing plant, the design summary could consist of a note stating the manufacturer's name, the make, and the size of the pump you are specifying. If the design summary is a note or a numerical value, it is often underlined or enclosed in a box on the page. The clearer you can make your thinking processes understood throughout the document, through the use of notes and figures, the more easily you or another engineer will be able to follow your calculations at a later date.

Computer results. Engineering analysis software can often be used not only to perform calculations, but also to produce a document that looks like engineering design sheets. The only special care you must take with these software packages is to ensure that your computer-generated document complies with the accepted guidelines for calculation documentation as described previously in this section. Use of other types of software (e.g., spreadsheets, finite element packages, or programs you write) requires special documentation.

Let's consider spreadsheets. The spreadsheet must be documented so that another individual working with a copy of your electronic file can understand your work. In your documentation, you should include at least one sample calculation, separate from the spreadsheet, to show how you set up the spreadsheet. A sample calculation or two is also important when you check the results of your computer analysis. In this case, your final calculation documentation might consist of a written document describing your spreadsheet solution, along with figures, notes, standards citations, and other items as appropriate, such as an attached electronic spreadsheet file. The spreadsheet must also be set up so that the user can understand how to use it. The explanation of the spreadsheet should be in standard engineering calculation format, including equations with defined variables. The spreadsheet itself, however, will be laid out with equations in terms of cell addresses rather than variables. If other people will use your spreadsheet or you expect to use it in the future, at a minimum you will need to define each variable, possibly in the cell above where the numerical value is to be placed. To make the spreadsheet even more transparent, you will also need to describe what the calculation result is from each cell, in a cell or cells near that result, and place a heading on all lists of data. You may even need to show the equation used in a cell, in terms of standard variable names, in a cell or cells near the cell that shows the result of using the equation. Complete and usable documentation of spreadsheets requires thought and care. Most spreadsheets are documented poorly. If you need to make a spreadsheet that must be well documented for ease of use and understanding, you should discuss the process with your colleagues and consider the preceding suggestions. Consider giving the spreadsheet to a colleague to see if she can use it without your help.

In dealing with software like finite element packages, you must address two issues when creating your design documentation. First, you must include enough information about your analysis (e.g., applied loads and boundary conditions, basic element sizes) that someone else could repeat your analysis by using the information in your documentation alone. Such usability could be accomplished by including figures and property data in your document, or the input files. If you are using finite element software that has extensive preprocessing capabilities, you will need to think carefully about how to include the appropriate information in your document. Second, in addition to clearly showing your modeling considerations, you must clearly show the salient results. Doing so can be accomplished by highlighting data from output files and referring to them at the appropriate place in your calculation documentation. If you are using software with significant postprocessing capabilities, you will probably want to include figures that graphically depict the analysis results. However, even if you include postprocessing figures, you will still need to document the sources of specific numbers used in your calculations.

If you use software you wrote or in-house software in your design analysis, you must include *well-commented* source code or describe in detail what the program does, with sample calculations included, similar to the requirements for spreadsheet documentation. Proprietary in-house software will need to be handled the latter way because you will not be able to include a listing of the source code with your document. Sample calculations for in-house software may not be necessary if you are working with a client who is familiar with the software. As always, follow the direction of your supervisor or instructor when creating documentation in this situation. Poorly documented and well-documented Java source code is shown in Figures 7.5a and 7.5b, respectively.

Checking and Double-Checking Your Calculations. Although checking your calculations is not part of the design documentation per se, it is an important part of the engineering process and worthy of mention. In some engineering offices, particularly structural engineering or civil works offices, another engineer, usually a registered professional engineer, always checks design calculations. In some cases, the calculations may also be checked by a different engineering firm. If you are an engineering student, your instructor or teaching assistant will thoroughly check your calculations. The fact that others will be checking your calculations clearly demonstrates the need for your documentation to be complete and understandable. Regardless of whether your calculations are checked by another person, you will need to perform some basic checks of the calculations to minimize the possibility of error. Ultimately, *you* are responsible for the quality of your work. Some suggestions for checking your calculations follow:

■ ***Review unit conversions.*** Incorrectly converting units will probably be the single most common type of error you will make when you perform your calculations. Remember the NASA Mars exploration vehicle that crashed probably because of a unit conversion error.

■ ***Make sure your results are reasonable.*** Look at the results of all equation calculations to ensure they are in the correct "ballpark." For instance, if an automobile crankshaft has been calculated to be 10 meters in diameter, this value is almost certainly incorrect. "Ballparking" your results can help you find misplaced decimals, units errors, or errors such as punching the wrong number on your calculator.

■ ***For complicated analyses, such as finite element analyses, try to determine some approximate solutions to the problem to see whether your answers are close to what you think they should be.*** You can also think of ways to perform simple checks using the analysis results. For example, in a structural analysis of a large building, you could make sure the given external loads on the overall building and the calculated reactions at the foundations satisfy the equations of static equilibrium.

■ ***Recalculate as many of your calculations as time permits.*** Recalculation is tedious, but it can help you find errors that might not be apparent when you are using the methods described previously.

7.3   Examples

Four examples of calculation documents and one example of source code commenting are presented in this section. Example 1 illustrates hand calculations on a standardized calculation sheet, Example 2 shows hand calculations on engineer's paper, Example 3 shows calculations performed and resulting documentation obtained from using engineering analysis software, and Example 4 shows calculations that were compiled on a word processor with embedded figures. Example 5 shows a good example and a poor example of source code commenting. Each example is discussed in detail next.

Example 1

Figure 7.1 shows the documentation of calculations performed for the design of reinforcement of an opening in a spherical shell. In this example, the calculations for this design solution were performed by hand; however, calculation sheets in electronic format in which you fill in the numbers and the computer does the math are currently available and used in some companies. Use of standardized sheets such as this makes the calculation documentation easier, but sketches, such as the one shown in Figure 7.1, would almost certainly need to be done by hand.

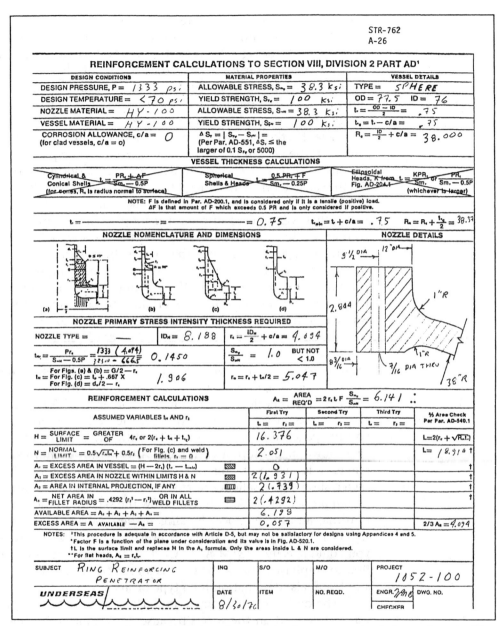

Figure 7.1. Example of calculation documentation on standard sheets (used with permission of Perry Slingsby Systems, Inc.)

Now, let's consider the content of the example document. Starting at the top of the page, STR-762 is the in-house number of the design report and A-26 is the page number for this sheet. The title tells you these are reinforcement calculations made in accordance with Section VIII, Division

2, Part AD of the ASME Pressure Vessel Code. The calculation sheet shown is an in-house tool, and a general assumption is that all users of the sheet will know, without being told, that the calculations were made in accordance with the ASME Pressure Vessel Code. Following the title are design conditions (criteria), material properties, and vessel details. The thickness calculations were crossed out because the thickness was determined earlier in the design report with a different technique.

The sheet uses drawings to show nomenclature and dimension variables (although they might not be entirely visible in this reproduction). The details for this particular nozzle are shown on a sketch drawn by the engineer who did the calculations. The actual calculations follow. Note that a significant portion of the sheet, at least half of it, is taken up by information related to the values that will be used in the calculations. Just performing the calculations without writing all this information on the sheet would be poor engineering practice. Last, at the bottom of the sheet is the company's name, the subject of the design, the project number, the date, and the initials of the engineer who performed the calculations. In this example, the design summary consists of the sketch showing the nozzle details (i.e., the objective of the design exercise was to determine the dimensions required for the nozzle).

___

Example 2

Example 2, shown in Figure 7.2, illustrates the documentation of calculations for the design of an attachment to a cylindrical shell. This document consists of hand calculations completed on engineer's paper.

Starting at the top of each page, the design report number, STR-762, and the page number are shown. The title, date, and engineer's initials are also shown at the top of each page. After the title information, the design reference is shown, in this case Welding Research Council Bulletin #107. A sketch showing all salient dimensions is shown next. The material types and design properties, S_m, are shown below the sketch, to the right. The assumption that another engineer looking at these calculations would know what A333 CL 1 means is reasonable. However, if you have any doubt about this, you could specify "ASTM A333 CL 1 steel" on the calculation documentation to emphasize that A333 CL 1 is an ASTM (American Society for Testing and Materials) designation for a type of steel.

The design calculations follow. The notation used in the calculations was not defined on the sheets because it all comes from the reference or is standard notation to a structural engineer or stress analyst. On the first page of the calculations, an approximation is used, and this fact is clearly stated. Whenever you do anything unusual in a set of calculations, you must clearly state what you are doing and, if appropriate, why. Notice on the second page of the document that tables and figures that were used from the reference publication are specifically noted, information that would be critically

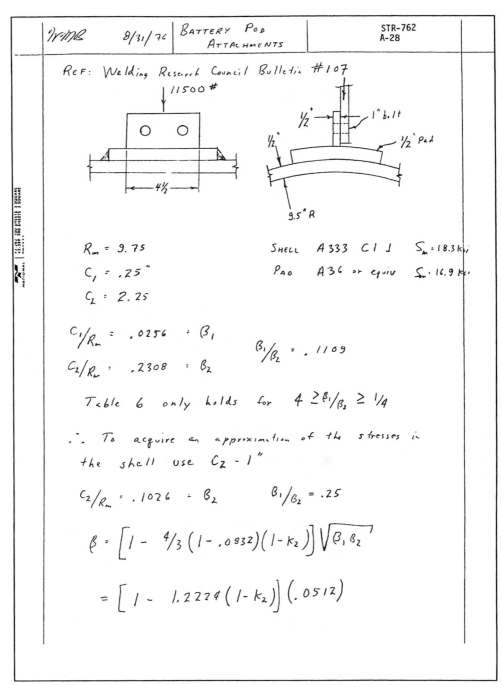

Figure 7.2. Example of calculation documentation on engineer's paper (used with permission of Perry Slingsby Systems, Inc.)

WMB   8/31/76

K_2 values taken from Table 6 WRC-107

N_ϕ $K_2 = 1.48$ $\beta = .0812$

M_ϕ $K_2 = .88$ $\beta = .0437$

N_x $K_2 = 1.2$ $\beta = .0637$

M_x $K_2 = 1.25$ $\beta = .0668$

Considering $\frac{1}{2}''$ thick pad

$T = 1.0''$ $\gamma = 10$

$R_m = 10''$

$M_\phi/P = .25$ (transverse axis) (Fig 1C)

$M_\phi/P = .27$ (long axis) (Fig. 1C-1)

$M_x/P = .195$ (trans. axis) (Fig. 2C)

$M_x/P = .195$ (long axis) (Fig. 2C-1)

$N_\phi/(P/R_m) = 1.8$ (Fig 3C)

$N_x/(P/R_m) = 2.0$ (Fig 4C)

Maximum Circumferential stress:

$\dfrac{N_\phi}{T} = (1.8)\dfrac{(11500)}{(10)(1.0)} = 2070$ ps. $\Rightarrow$ Max $\sigma_\phi = \pm 20700$ psi

$\dfrac{6M_\phi}{T^2} = (.27)\dfrac{6(11500)}{(1)^2} = 18630$ psi

Maximum Longitudinal Stress:

$\dfrac{N_x}{T} = (2.0)\dfrac{(11500)}{(10)(1.0)} = 2300$ psi $\Rightarrow$ Max $\sigma_x = \pm 15755$ ps

$\dfrac{6M_x}{} = (.195)\dfrac{(6)(11500)}{} = 13455$ psi

Figure 7.2. Example of calculation documentation on engineer's paper (*continued*)

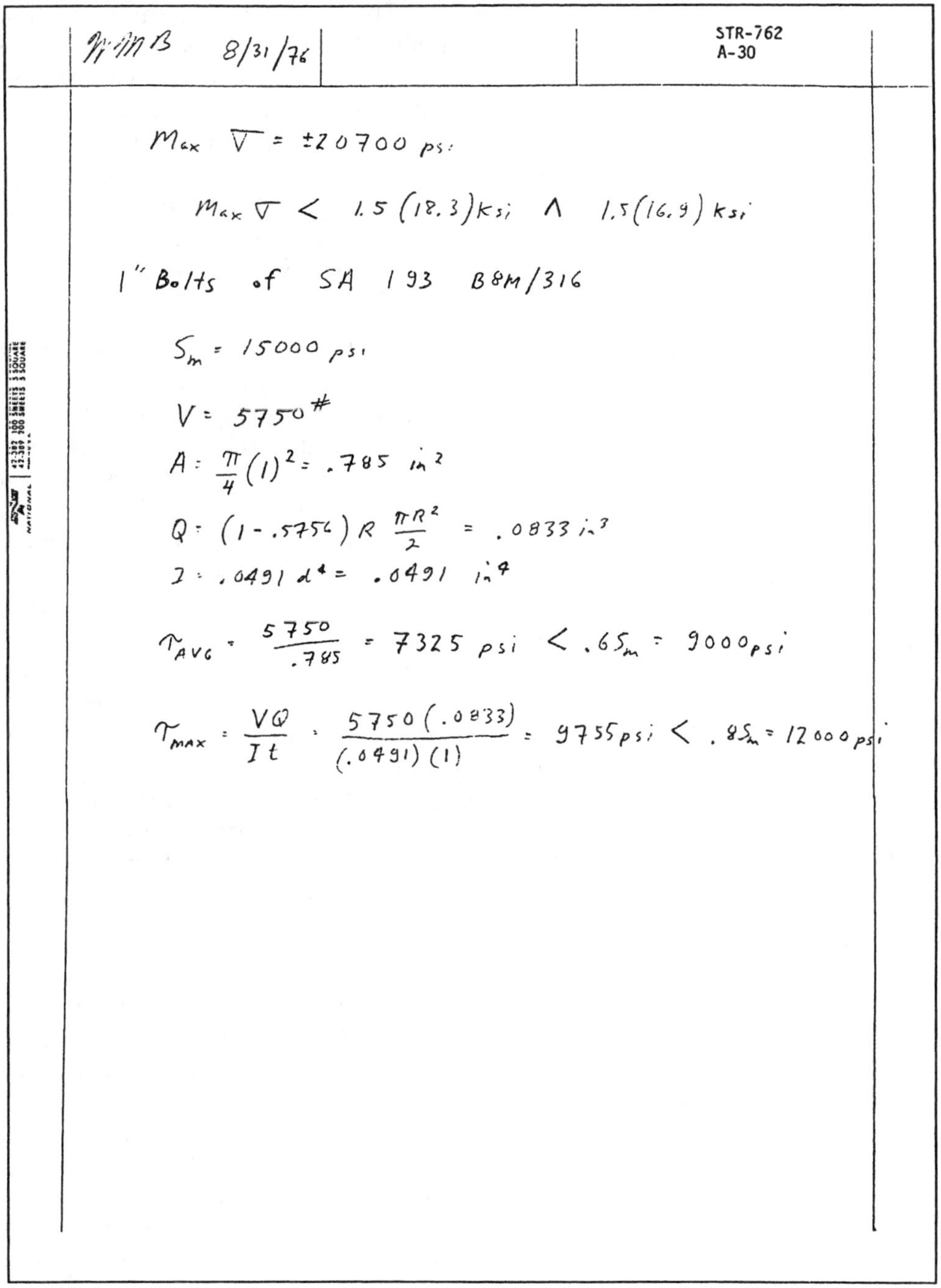

WMB 8/31/76 | STR-762 A-30

Max $\sigma = \pm 20700$ psi

Max $\sigma < 1.5(18.3)$ ksi $\wedge$ $1.5(16.9)$ ksi

1" Bolts of SA 193 B8M/316

$S_m = 15000$ psi

$V = 5750\#$

$A = \frac{\pi}{4}(1)^2 = .785 \text{ in}^2$

$Q = (1 - .5756) R \frac{\pi R^2}{2} = .0833 \text{ in}^3$

$I = .0491 d^4 = .0491 \text{ in}^4$

$\tau_{AVG} = \frac{5750}{.785} = 7325 \text{ psi} < .6 S_m = 9000 \text{ psi}$

$\tau_{MAX} = \frac{VQ}{It} = \frac{5750(.0833)}{(.0491)(1)} = 9755 \text{ psi} < .8 S_m = 12000 \text{ psi}$

Figure 7.2. Example of calculation documentation on engineer's paper (*continued*) important to anyone else who was examining the calculations. On the third page of the document, all calculated stresses are shown to be less than the allowable values. Showing that the calculated stresses are less than the allowable stresses is the design summary for this particular set of calculations.

Another way to create this document would have been to list all the allowable stresses at the beginning of the calculations. Regardless of where the allowable stresses are listed, you should still show clearly, at the end of the document, that the design criteria have been met (e.g., that the actual stress is less than the maximum allowable stress).

Example 3

Example 3, shown in Figure 7.3, is the documentation for the design of a steel angle in tension. This example was completed with computer analysis software. This particular design required no figures, but if figures had been necessary, they could have been included on the document as hand-drawn sketches because hand drawing is still generally the fastest way to include them in the document. Alternatively, the figures could have been created with some type of drawing tool and then imported into the document. Some analysis tools do not have a good built-in drawing package, and importing a drawing from another package may result in extremely large electronic files. In either case, a hand-drawn sketch on the document would probably be best.

The design specification that was followed is stated at the top of the document. In this example, only one specification was used, so any reference to a section or table must refer to the stated specification. The material, member length, and loads (i.e., the given information) are stated at the beginning of the design calculation document. The design follows the stated given information, with explanatory notes and design specification references included as needed. A summary statement, rather than a summary sketch, is given at the end of the document. The summary is emphasized by using bold letters in a larger font size.

Example 4

Example 4, shown in Figure 7.4, shows the determination of the line resistance and line current for a circuit. The circuit is shown in a figure, followed by the given resistor information. The analysis is performed step by step, group by group for ease of understanding. An intermediate step in the analysis is shown by the second figure to help explain the remainder of the calculations. A summary of the analysis is shown at the end. This set of calculations was performed on separate sheets of paper and then transferred into its present form by using a word processor. The sketches were drawn with a drawing package then imported into the word processor document. This approach to calculation documentation is not common in engineering design, but it is a possibility if you want your design to be neatly done and include embedded figures.

Design of a Steel Angle in Tension

Select the lightest single angle. Design to AISC, *Manual of Steel Construction,*
Load and Resistance Factor Design, 3rd ed.

Use A36 steel

Member tension loads:

Dead load: $\quad P_D := 10 \quad$ kips
Live load: $\quad P_L := 45 \quad$ kips

Material Properties:

Yield stress $\qquad F_y := 36 \quad$ ksi
Ultimate stress $\quad F_u := 58 \quad$ ksi

The member is 15 feet long. $\qquad L := 12 \cdot 15 \quad$ in.

The member will be connected to a gusset plate by using a single-gage line of at least
3, 3/4″-diameter bolts.

$\qquad U := 0.85 \qquad$ Commentary, Section B3(b)

Calculate factored axial load:

$\qquad P_u := 1.2 \cdot P_D + 1.6 \cdot P_L$
$\qquad P_u = 84 \quad$ kips

The design will be controlled by yielding on the gross section or fracture on the net section.

Design such that Pu $<. \phi_t P_n$

For yield:

$\qquad \phi_t P_n = 0.9 A_g F_y \qquad\qquad$ Section D1(a)

thus the minimum A_g is: $\quad A_{gmin} := \dfrac{P_u}{0.9 \cdot F_y} \qquad A_{gmin} = 2.593 \qquad$ in.^{2}

For fracture:

$\qquad \phi_t P_n = 0.75 U A_n F_u \qquad$ Section D1(b)

thus the minimum A_n is: $\quad A_{nmin} := \dfrac{P_u}{0.75 \cdot U \cdot F_u} \qquad A_{nmin} = 2.272 \qquad$ in.^{2}

Effective hole diameter: D = 3/4 − 1/16 − 1/16 = 7/8 inch. Table J3.3 & Section B2.

$\qquad D := 0.875 \qquad$ in.

Minimum radius of gyration as controlled by minimum slenderness ratio. Section B7.
Note: The minimum radius of gyration for an angle is about the z axis.

$\qquad r_{min} := \dfrac{L}{300} \qquad$ thus $\qquad r_{min} = 0.6$

Considering the minimum gross area, A_{gmin}, try a 4 × 3 1/2 × 3/8 angle with a cross-sectional area of 2.68 in.^{2} and a weight of 9.1 lb/ft. Table 1.7

$\qquad A_g = 2.68$ in.$^2 > A_{gmin} = 2.593$ in.^{2}

$\qquad r_z = 0.719 > r_{min} = 0.60$

$\qquad t := 0.375 \qquad$ in.

$\qquad A_n := 2.68 - D \cdot t$

$\qquad A_n = 2.352 \quad$ in.$^2 > A_{nmin} = 2.272$ in.^{2}

This is the lightest angle in Table 1.7 that meets the design criteria.

Use a 4 × 3 1/2 × 3/8 angle of A36 steel.

Figure 7.3. Example using design calculation software

Direct Current Circuit Analysis

Find the total resistance and line current for the circuit and corresponding resistances shown next.

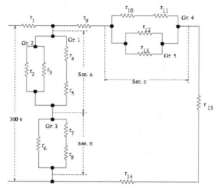

The resistances (in ohms) are: $r_1 = 30$, $r_2 = 800$, $r_3 = 1200$, $r_4 = 300$, $r_5 = 400$, $r_6 = 800$, $r_7 = 800$, $r_8 = 300$, $r_9 = 20$, $r_{10} = 200$, $r_{11} = 100$, $r_{12} = 80$, $r_{13} = 200$, $r_{14} = 30$, and $r_{15} = 40$.

Group 1: Using the rule for resistors in series:

$$R_1 = r_4 + r_5 = 300 + 400 = 700$$

Group 2: Using the rule for resistors in parallel:

$$1/R_2 = 1/r_2 + 1/r_3 = 1/800 + 1/1200 \Rightarrow R_2 = 480$$

Section a:

$$1/R_a = 1/R_1 + 1/R_2 = 1/700 + 1/480 \Rightarrow R_a = 285$$

Group 3:

$$R_3 = r_7 + r_8 = 800 + 300 = 1100$$

Section b:

$$1/R_b = 1/r_6 + 1/R_3 = 1/800 + 1/1100 \Rightarrow R_b = 463$$

Group 4:

$$R_4 = r_{10} + r_{11} = 200 + 100 = 300$$

Group 5:

$$1/R_5 = 1/r_{12} = 1/r_{13} = 1/80 + 1/200 \Rightarrow R_5 = 57$$

Section c:

$$1/R_c = 1/R_4 + 1/R_5 = 1/300 + 1/57 \Rightarrow R_c = 48$$

The reduced circuit, consisting of the remaining actual resistors and the equivalent resistors, is shown next:

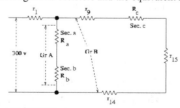

Figure 7.4. Circuit analysis

Group A:

$R_A = R_a + R_b = 285 + 463 = 748$

Group B:

$R_B = r_9 + R_c + r_{14} + r_{15} = 20 + 48 + 30 + 40 = 138$

Groups A and B are in parallel, thus

$1/R_{AB} = 1/R_A + 1/R_B = 1/748 + 1/138 \Rightarrow R_{AB} = 117$

Therefore, the total resistance is:

$R_T = r_1 + R_{AB} = 30 + 117 = 147$ ohms

And, the line current is:

$1 = V/R_T = 300/147 = 2.04$ amps

Summary

Total resistance: 147 ohms
Line current: 2.0 amps

Figure 7.4. Circuit analysis (*continued*)

Example 5

Example 5, shown in Figures 7.5a and 7.5b, is a Java program that finds the real roots of a quadratic equation. Figure 7.5a shows poorly documented source code, and Figure 7.5b shows a well-documented version. Figure 7.5b shows judicious use of white space as well. The program in Figure 7.5a performs the same operations as that in Figure 7.5b, but the well-documented code with good use of white space is much easier for a person to read.

```
import javax.swing.*;
public class Quadratic {
public static void main(String args[])
{
String s1=JOptionPane.showInputDialog(
"Enter A");
String s2=JOptionPane.showInputDialog(
"Enter B");
String s3=JOptionPane.showInputDialog(
"Enter C");
double A = Double.parseDouble(s1);
double B = Double.parseDouble(s2);
double C = Double.parseDouble(s3);
double D = B*B-4.0*A*C;

if (D>0.0) {
double X1 = (Math.sqrt(D)-B)/(2.0*A);
double X2 = (-(Math.sqrt(D))-B)/(2.0*A);
JOptionPane.showMessageDialog(null, "A: " + A + "\nB: " +
B + "\nC: " + C + "\nFirst Root: " + X1 +
"\nSecond Root: " + X2, "Real Roots", JOptionPane.PLAIN_MESSAGE);
}
else if(D<0.0)
JOptionPane.showMessageDialog(null, "No Real Roots: ",
"Real Roots", JOptionPane.PLAIN_MESSAGE);
else {
double X = -B/(2.0*A);
JOptionPane.showMessageDialog(null, "A: " + A + "\nB: " +
B + "\nC: " + C +"\nOnly Real Root: " + X,
"Real Roots", JOptionPane.PLAIN_MESSAGE);
}
System.exit(0);
}
}
```

No comments to explain program logic

Poor use of white space

Figure 7.5A. Example of a poorly documented source code

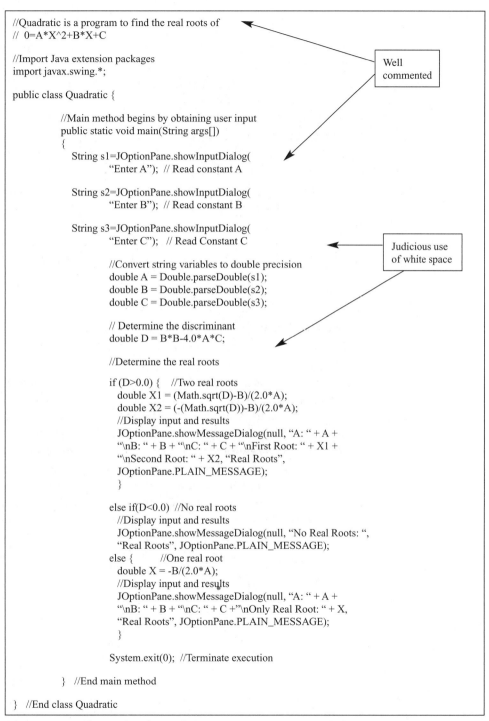

```
//Quadratic is a program to find the real roots of
//  0=A*X^2+B*X+C

//Import Java extension packages
import javax.swing.*;

public class Quadratic {

        //Main method begins by obtaining user input
        public static void main(String args[])
        {
            String s1=JOptionPane.showInputDialog(
                    "Enter A");  // Read constant A

            String s2=JOptionPane.showInputDialog(
                    "Enter B");  // Read constant B

            String s3=JOptionPane.showInputDialog(
                    "Enter C");   // Read Constant C

                //Convert string variables to double precision
                double A = Double.parseDouble(s1);
                double B = Double.parseDouble(s2);
                double C = Double.parseDouble(s3);

                // Determine the discriminant
                double D = B*B-4.0*A*C;

                //Determine the real roots

                if (D>0.0) {    //Two real roots
                  double X1 = (Math.sqrt(D)-B)/(2.0*A);
                  double X2 = (-(Math.sqrt(D))-B)/(2.0*A);
                  //Display input and results
                  JOptionPane.showMessageDialog(null, "A: " + A +
                  "\nB: " + B + "\nC: " + C + "\nFirst Root: " + X1 +
                  "\nSecond Root: " + X2, "Real Roots",
                  JOptionPane.PLAIN_MESSAGE);
                  }

                else if(D<0.0)  //No real roots
                  //Display input and results
                  JOptionPane.showMessageDialog(null, "No Real Roots: ",
                  "Real Roots", JOptionPane.PLAIN_MESSAGE);
                else {          //One real root
                  double X = -B/(2.0*A);
                  //Display input and results
                  JOptionPane.showMessageDialog(null, "A: " + A +
                  "\nB: " + B + "\nC: " + C +"\nOnly Real Root: " + X,
                  "Real Roots", JOptionPane.PLAIN_MESSAGE);
                  }

                System.exit(0); //Terminate execution

        }  //End main method

}  //End class Quadratic
```

Well commented

Judicious use of white space

Figure 7.5B. Example of a well documented source code

7.4 Exercises

1. Create a checklist for calculation documents that you can use in your future engineering classes.

2. Create a calculation document, on engineer's paper, for a problem assigned in one of your engineering courses.

3. Create a calculation document, on engineer's paper, that answers the following question: How many toothpicks can be made from a log measuring 3 feet in diameter by 20 feet long? Include all assumptions in your document.

4. Create a calculation document, on engineer's paper, that estimates the maximum number of cars per hour that can travel down two lanes of an expressway as a function of time. For safety reasons, cars traveling at 60 miles per hour should be spaced at least six car lengths apart.

5. Create a calculation document, on engineer's paper, that estimates household energy expenses for a year. Electricity (for heat, air-conditioning, and general use) costs $0.07 per kilowatt hour, and gasoline costs $1.69 per gallon. Include all assumptions about climate, house size, distance from work, and so forth in your document.

6. Do Exercise 2 using engineering analysis software.

7. Do Exercise 3, 4, or 5 using engineering analysis software.

8. Exchange a copy of your document from Exercise 2 or 6 for a classmate's document for the same exercise. Each of you should review the other's document to determine whether it is complete, making notes directly on the document. Because you both did the same problem, you must be careful to review the document as if you were seeing the work for the first time. Hand in the reviewed documents.

9. Do Exercise 4 and exchange a copy of it for a copy of another classmate's work on Exercise 5. Each of you should review the other's document to determine whether it is complete, making notes directly on the document. Hand in the reviewed documents.

8

Oral Communication

Sometime in your career, both academic and professional, you will need to give a presentation. The presentation may be given to a range of possible audiences, as discussed in the audience analysis section of Chapter 2. Whatever the audience, you will need to choose the tools and techniques appropriate for this type of communication. In this chapter, we help you with these decisions and give you some pointers on how to create and give a high-quality presentation. When you finish this chapter, you may want further information on giving presentations. Thus, we included a few additional resources in Section 8.8 to help you locate such information.

8.1 Tool Selection for Technical Presentations

The type of tool you choose to use for oral presentations will depend somewhat on the setting in which you will be speaking. For example, if you were making a professional presentation at a conference, you would likely choose presentation software to use to prepare your visual aids for your speech. At a conference, you would have a reasonable expectation that a screen, a computer projection system, and possibly an audio system would be available to you for your presentation. If you were speaking in front of a fifth-grade class, you would probably want to limit your visual aids to either overheads or slides (arranging this setup with the teacher beforehand) or to photographs you could pass around the room for examination. If you were speaking in front of a town council, you would want to check with council leadership in advance about the ability to use visual aids during presentations, but you might be limited to tools like flip charts in this instance.

If you do not have presentation software available for making overheads for presentations, you can use a word processor to create no-frills visual aids. Just be sure to set your font sizes large enough that the writing on the overheads is visible from the back of the room in which you will be speaking. A reasonable rule to follow is to place the overheads at your feet and stand up. If you can read them easily from this vantage point, they will project large enough on the screen so that most attendees will be able to see them. This technique will work for small to moderate-size rooms with standard-size screens, the kind you will encounter most often. For large rooms that seat 200 people or more, you will need to consider the screen size and the maximum distance from any audience member to the screen. A slide no more complicated than Figure 8.2a is a good target.

8.2 Types of Presentations

Presentations can be categorized a number of ways. The first category is *formal or informal*. We emphasize formal presentations; informal presentations are fairly straightforward once you understand formal presentations. The second category is *individual or group* presentations. We emphasize individual presentations; group presentations follow from that. The third category concerns the possible *kinds* of presentations you might need to make. This category includes technical papers, informational presentations, sales presentations, progress reports, project reports, tutorials, and proposals. The kind of presentation usually indicates the type of audience you will be facing. Once you know the kind of presentation you will be giving, the next step is to do some audience analysis, as discussed in Chapter 2.

Whether you will be giving a formal or an informal presentation as an individual or in a group is usually decided for you. The decision is based on the circumstances surrounding the presentation. If you're presenting a design report to a client and the entire design team is present, you will likely give a formal group presentation, particularly if each team member had a distinct responsibility in the design. If you're presenting a technical paper at a conference, you will most likely give a formal individual presentation. However, if you're giving an informational presentation to a community group, you will need to decide how formal or informal the presentation should be. In this case, you should ask one of the group leaders what type of presentation the group would like. If you can't get a clear answer, err on the side of more formal. Note that *formal* does not mean stilted. Whether a presentation is formal or informal, it should be interesting and enjoyable to listen to.

8.3 Formal Individual Presentations

The formal individual presentation is the basic type of presentation; it is also the most common type you will encounter as an engineer. As stated previously, if you can give this type of presentation well, the other types will follow.

Function. Presentations are a way to deliver information in a personal and, if done well, entertaining manner. Formal presentations (like all presentations) allow you to control carefully how the information you want to get across is delivered to the audience. This type of control means you can emphasize what you want the audience to take away from the presentation. Such control is much more difficult to achieve in written communication. However, a formal presentation also requires that you carefully control how the information is delivered. Formal oral presentations are a double-edged sword. They allow you the control, but they also demand that you take control. You, as the presenter, control the order in which the information is presented. The listener cannot skip around as a reader of written communication can.

Generally, presentations have a relatively short time limit and are given to a group. Thus, you can present information to more people in a shorter time than you can with written communication. However, since the time limit is short, you must make every minute count. You must present as much information as you can, but not so much that you overload the audience. If you overload the audience, you will lose its attention, and all your efforts will be of no value. In the next two sections, we present the form and the mechanics of a formal presentation so that you can deliver an organized, effective presentation.

Form. The basic components of a presentation (formal or otherwise) are the introduction, the body, and the ending. Let's consider each.

Introduction. The introduction should include the title of the presentation, the name of the author or authors, a statement about why the presentation is important to the audience, and a brief overview of the remainder of the presentation. You should also use the introduction to gain rapport with the audience. A pleasant attitude and use of the word *you* will help.

Body. The body is the majority of the presentation. In this section, you present the information in a logical, clear fashion. This material must be well connected so that the audience can follow your train of thought. The body may require subsections. If so, the subsections should cover major ideas you want to present. Each subsection should consist of explanations that help the audience understand the major idea of that subsection. These explanations should tie technical and unfamiliar information to information that most of the audience is familiar with. The subsections need to be connected to each other, and the information within each subsection needs to be connected. The interconnectedness of a presentation is vital to holding the audience's attention and aiding understanding. Remember that an audience member cannot go back and "reread a paragraph."

Ending. The ending should include a summary or an overview of the material presented. It should also include recommendations if you are presenting a project, research, or a design report. Let the audience know you are ending by beginning the ending with words such as "Today I've discussed," "In summary," or "In conclusion." After you complete the ending, ask whether the audience has any questions.

Mechanics. We now discuss the details of designing and presenting an effective presentation.

> ***Organizing your presentation.*** A good presentation must be well organized. Generally, the audience depends entirely on you for all the information needed to understand your presentation. We suggest you outline the presentation. Outlining was not emphasized for technical writing because many successful writers do not outline; it is a matter of personal preference. The reason we suggest you outline a presentation is that you must make every word count. Furthermore, if you plan to use visual aids such as presentation software or transparencies, the items in your outline will correspond almost exactly to the *slides* you will show. (The word *slide* is used to refer to the image you show the audience. A slide could be an overhead transparency, a photograph, or a screen from presentation software.) If you are giving a presentation that does not require visual aids, you should make note cards from your outline. Each note card should have a few keywords on it that will prompt you about what to say next. Do not put the entire presentation on the note cards. The basic form for a presentation was discussed previously; how you choose to design your presentation within these guidelines is your decision.

> ***Developing your slides.*** We assume in this section that you will use visual aids. In formal presentations, particularly those that are technical, visuals are vital. On the one hand, most people have difficulty absorbing technical information simply by listening. On the other hand, most people also have difficulty absorbing large amounts of visual information. You must strike a balance.

> We consider four common types of slides: pictures or diagrams, graphs or charts, tables, and words. First, regardless of the type of slide, make sure the slide is needed. You must have a reason for using each slide and something to say about it. If you have nothing significant to say about a slide, omit it. Also, if you want to use a specific slide more than once, put it in the presentation at each point required. Do not search back and forth to find it since doing so will be too distracting for both you and the audience. Second, the information on a slide must be readable by everyone in the audience. This guideline means pictures and diagrams must be large and not too detailed, and graphs and charts must be large and not too cluttered. We suggest no more than about five items on a graph or chart (e.g., lines or bars). Pie charts may have more than this, but still be careful about too much detail. *Remember:* You are explaining the picture,

diagram, graph, or chart, so make the labels simple and big. Consider the examples shown in Section 8.6.

If you plan to use tables on your slides, design them for the audience. Tables appropriate for a report or paper are generally not suitable for a slide. A slide should have no more than 5 to 10 lines on it. Thus, a table with many rows and columns will be too detailed for the audience. You should restrict any table to a maximum of 5 to 10 rows and about 5 columns. Even at this size, you will need to point out which table term you are discussing as you discuss it.

Slides with words on them will likely be a significant portion of your presentation. As for tables, use no more than 5 to 10 lines on a slide. Use a large font to be sure everyone in the audience can read the slide; we suggest 24-point type. Also, use an easy-to-read font, such as Helvetica. Use capital and lowercase letters. Do not use *all* capital letters because using all caps makes the slide more difficult to read. Use a simple background. *Remember:* You do not want to distract the audience from what you are saying. Finally, for all types of slides, if you must say, "I know you can't read (see) this, but . . . " do not use the slide. What good is a slide if the audience can't read it?

If you are using presentation software, you probably have the ability to use various backgrounds, colored type, art (often called *clip art*), animation, and even digital video. Use a fairly subtle background; you want the audience to look at the words or figures, not the background. Use a font color that contrasts well with the background you choose, but not so bright or loud that it overpowers the audience. A *trial audience* (friends, relatives, etc.) can help you decide whether your background and font choices are appropriate. Art, animation, and digital video should be used sparingly and only when it will enhance the presentation. You want to "wow" the audience with your overall presentation, not with how accomplished you are with the presentation software. Remember that the audience can absorb only a limited amount of information in the time you have allotted for your presentation, and too much information will overwhelm audience members.

Making note cards. The guidelines for slides with words are also appropriate for note cards. Use no more than 5 to 10 lines on a note card. Five lines is probably best to make the cards easy to read. A good idea is to number the cards in the order in which you want to use them. This practice might sound silly, but if you accidentally shuffle the cards during your presentation, you will be better able to recover if the cards are clearly numbered. Under no conditions should the entire presentation be written on note cards. Some people use note cards even when they have visuals, but we recommend that your visual aids be designed so that they can be used in place of note cards.

Practicing your presentation. Be sure to practice your presentation. We recommend that you go through the full presentation at least twice. Practice will help you make sure you have included everything you want to include and help you discover items that can be left out. Typically, you should not memorize your presentation. Doing so can disconnect you from your audience, and, worse, if you forget a word or phrase, you may lose your entire train of thought. If, at the other extreme, you believe you need both note cards and visual aids, you should either redesign your visual aids or practice until you don't need the note cards.

Generally, you will have a time limit for your presentation. Practicing will also help you make sure the presentation is an appropriate length. If you are using slides, a guideline that has worked for us is 1 minute per slide. You might find this guideline unsuitable for you, but it seems to work for many engineers we know.

If you can give your presentation to a trial audience, do so. A trial audience can best tell you where you have problems in either content or delivery. We also suggest that, at the least, you go to the room in which the presentation will be given to "get a feel" for the place where you must speak. Actually standing where you will give the presentation (most presentations are given while you are standing) and going through it, verbally or mentally, is ideal.

Delivering your presentation. You can do a number of things to help yourself give a good presentation, and you should not do a number of other things. We discuss both in this section.

■ ***Accept nervousness.*** Almost everyone you ask will tell you they are nervous before a speech. Expect to be nervous, and live with it. Mark Twain called this type of nervousness "the sweat of perfection." One of the authors, who has been teaching for 20 years, is still nervous when he (now you know which one) enters a classroom to teach a class. The nervousness lasts until only the first few words are spoken, but exists nonetheless. To help control your nervousness, you can take a few deep breaths (try not to be too obvious) before you go forward to give your presentation. Another help is to start your presentation with a neutral, easy-to-say phrase (e.g., "Good morning" or "Good afternoon"). Using a neutral phrase allows you to "test" your voice and gets you started.

■ ***Make eye contact.*** You must look at the audience. Do not speak to the screen or look at the floor, the back wall, the ceiling, or anywhere other than at the audience. When you open with your starting phrase and your introduction, look around the room for friendly faces. You should be able to find a friendly face in multiple locations around the room. As you give your presentation, look around to each friendly face, and you will automatically be looking at the entire audience. The more people in the audience you can look at during your presentation, the better your presentation will be. In small groups, you should look at everyone. With large groups, you should look at as many people as possible, but at least enough so that you appear to be looking at the entire audience. The idea is to make each person in the audience think you are talking to her or him personally.

■ *Use appropriate body language.* Avoid being too stiff or too theatrical. For most people, their natural approach to speaking is most appropriate. Use gestures as suitable, but be cautious of mannerisms such as jingling the keys or the change in your pocket, keeping one hand in your pocket, waving the pointer around, or gesturing wildly. Search for a middle ground between no motion and frenetic behavior. If other speakers are using a lectern or the only microphone is at the lectern, use the lectern. When using a lectern, be careful not to lean on it. It is a place to set your notes, if you have any, but not much else should be placed there. Many people prefer to speak without a lectern. If you don't need a microphone (when in doubt use one) or a lapel microphone or another type of portable microphone is available and you can do without the lectern, don't use it. Standing away from the lectern will help you connect better with your audience: you will appear to be more informal and conversational. However, if you do not use a lectern, you must be careful of your body language since you will be completely visible. A hand in a pocket is much less apparent when you are behind a lectern than it is when you're standing away from it.

■ *Speak appropriately.* You should speak slowly and clearly. You will probably tend to speak too fast when giving a presentation, so if you concentrate on speaking slowly and clearly, you will reduce that tendency and be more easily understood. Speak loudly enough to be heard easily by the entire audience. When you start your presentation, ask the audience members whether they can hear you. If a microphone is available, use it. Most people speak too softly at times, and nothing ruins a presentation more than if the audience must strain to hear. Conversely, don't yell or set the microphone volume too high. The audience should be comfortable. Finally, control verbal mannerisms. Be careful about saying the phrases "You know" and "Okay." These phrases are never needed. Use the word *like* correctly. You do not need fillers such as *uh* and *um*. If you need to pause to think, just pause. The silence may feel interminable to you, but in most cases it won't to the audience, and silence is always better than *uhhhhhh*. So, you should *not* say, "You know, like use the word *like* correctly. Uhhhh, okay?"

■ *Use your slides correctly.* If you are using visuals, use them effectively. First, be careful not to stand in front of the screen. The slides will be effective only if the audience can see them. Remember "line of sight"; even if you're off to the side of the screen, you still might be blocking people on the edge of the audience. Second, explain each slide and use a pointer as necessary. You included each slide in the presentation for a reason, so explain the content of the slide. If you have nothing to say about a slide, it should not be in the presentation. Slides such as figures, graphs, charts, tables, and diagrams require explanation. You should "walk" the audience through each slide and use a pointer to direct attention to specific details you want to emphasize. Finally, do not read your slides. The audience can read the slides. Instead, fill in the gaps since the slides should be reminders of what to say, not the entire thought. If you use a quotation or two in the presentation and put them on a slide, you may read them if you believe doing so is necessary. Many speakers give the audience time to read quotations for itself. Either method is acceptable, but audience members can usually comprehend the slide better if they read it than if it is read to them.

■ *Help the audience.* You can help the audience understand your presentation by using techniques that guide the audience. One technique is to repeat your key points. Your presentation will have one or more key points or ideas that you want the audience to remember. Mention these points a few times during the presentation. You may want to emphasize that you are repeating a point by saying something such as, "As I said before. . . ." You should present conclusions at the end of your presentation, and they should be based on the ideas in your presentation: they are your final opportunity to repeat some of your main points. Another technique is to emphasize transitions in your presentation. Since listeners cannot reread something they missed, your train of thought must be extremely clear. You can emphasize the transitions by using words or phrases that show the audience when a transition is occurring. Examples of these phrases or words are *First, As a result of, In conclusion,* and *On the other hand.* The purpose of words and phrases such as these is to help the audience see the connections between the ideas you are discussing in your presentation.

■ *Stay within your time limit.* Your presentation will have a time limit. Do not exceed this limit. If you do and other speakers follow you, you will be taking time from them. If you speak beyond the time limit, you are implying that what you have to say is more important than what later speakers have to say. Such implied arrogance will give both the later speakers and the audience a bad impression of you. Even if you are the sole speaker, the people in the audience will likely have scheduled their day assuming you would finish on time. Again, exceeding the time limit implies that your presentation is more important than whatever else they have to do. Last, be sure to leave time for questions. A few people in the audience will likely want to ask questions.

■ *Be ready for questions.* You should expect questions. Prepare yourself for questions by thinking of questions *you* might have if you heard your presentation. You will not be able to think of all the possible questions, so be prepared to not know the answer to a question. In such a case, you may speculate if you like, but nothing is wrong with simply saying you don't know. You may also want to offer to talk to the questioner later. This reply is appropriate if the answer would take too long, if the question or questioner is hostile, or if the question is not related to your topic. For most presentations, saving the questions until the end is best for the speaker, particularly if the time limit is rigid. If the time limit is flexible and you think you can handle questions during your presentation without losing your train of thought, allow questions as you go. This format is best for the audience. However, if you choose this approach, be prepared to revert to questions at the end if you feel as if you are losing control of your presentation.

8.4 Formal Group Presentations

Since engineers often work on projects as a team, you will encounter formal group presentations fairly often. Such presentations have much in common with formal individual presentations, so we primarily discuss the differences.

Function. The functions described for formal individual presentations also apply to formal group presentations. However, group presentations have a few more functions. For group presentations, the person with the most knowledge about a particular portion of a project should be the speaker. Taking this approach should mean that the explanation of the topic is better. The variety of speakers will likely help the audience stay focused on the presentation, and no individual presenter's workload will be too great. Group presentations are often longer than individual presentations, and a team will generally be able to speak longer than a single individual.

Form. The form is the same as for a formal individual presentation. The difference is simply that the group members will divide the presentation in some manner.

Mechanics. All the mechanics discussed for a formal individual presentation apply to a group presentation. However, some additional mechanics also need to be considered.

Organizing the presentation. The only additional organization required is the division of the presentation so that each portion can be allocated to a group member.

Developing your slides. Since the presentation must look like one presentation, the group members must work together to make sure each portion of the presentation looks like all the others. Layouts, fonts, and styles must all be the same. If each portion is different, the audience will be distracted and have more difficulty following the presentation.

Practicing your presentation. Again, practice is important. However, for a group presentation, the group must practice together. Each person should practice his portion, but the group must also practice together to make sure the entire presentation is not disjointed.

Delivering your presentation. All the mechanics discussed for delivering a formal individual presentation apply. However, a few additional mechanics are necessary for a group presentation.

- ***Introduce the group.*** You should introduce all the group members and each of their responsibilities at the beginning of the presentation. Showing a slide containing this information is a good idea.
- ***Work on transitions.*** The transition from speaker to speaker should be smooth. Each speaker should introduce the next speaker. A lead-in such as ". . . and now Diane will discuss the pump design" is one approach to this type of transition.
- ***Refer to previous speakers.*** Your group presentation will be better if speakers can weave previous speakers' information into their portion of the presentation. Doing so requires practice.
- ***Dress consistently.*** Each group member should be dressed at about the same level of formality. One member in a full three-piece suit and another in jeans and a T-shirt will be disconcerting to the audience and indicate the group is not working together. Each member does not need to be dressed the same as each other member, but similar attire will make the presentation look more like a group effort.
- ***Stay within your time limit.*** Not only should the entire presentation stay within the time limit, but so should each speaker's portion. Exceeding the time allotted to your portion of the presentation is unfair to your team members. If the entire presentation has a hard time limit and you exceed your time allotment, the end of the presentation, when the important conclusions and recommendations are given, may have to be omitted. So, stay within your time allotment. Doing so requires practice.
- ***Be ready for questions.*** Your group must decide whether questions will be allowed for each speaker or will be held until the end of the presentation. If the overall time limit for the presentation is fairly flexible, allowing the audience to ask each speaker questions during or at the

end of her portion is the best format for the audience. However, be prepared to revert to questions at the end if you think you are losing control of the overall presentation. Questions at the end are easier for the group, particularly if the time limit is rigid. In either case, let the group member who is best prepared to answer a question do so. Within reason, group members may expand on or clarify answers given by another group member, but if such comments are made too often, they may indicate the group is not working together. *Remember:* You are a team who supposedly agrees on whatever you are presenting.

8.5 Informal Presentations

The preparation for informal presentations—individual or group—is more limited than that for formal presentations.

Function. Informal presentations serve the same function as formal presentations, just in a less formal manner, which could mean, for instance, less preparation, a less formal setting, fewer and more quickly prepared visual aids, or no visuals. Whatever the limitations, the goal is the same: present the information to the audience in as clear and entertaining a manner as possible.

Form. Informal presentations are generally shorter than formal presentations. Even so, the presentation should include an introduction, a body, and an ending, just as for a formal presentation. The goal is to make an informal presentation as close to a formal presentation as possible within the limits placed on you. So, if the informal presentation is an impromptu speech in which you have little or no time to prepare, you should still think in terms of the three basic parts.

Mechanics. The mechanics of formal presentations should be applied to informal presentations as much as possible. Preparation of slides or note cards and practice of the presentation may be impossible, but the delivery mechanics apply. You might not have prepared slides, but you should use whatever tools are available (e.g., blackboards, whiteboards, or flip charts) to develop impromptu visuals. You should still accept your nervousness, make eye contact with the audience, use an informal manner, use appropriate language, watch out for mannerisms (both body language and verbal), and be ready for questions. Since the presentation is informal, you should have a relaxed manner but still not lean on the lectern, put your hands in your pockets, or slouch. One mistake students often make during informal presentations is to think they can speak as if they are talking to their friends at home. This misconception results in inappropriate language and mannerisms. A better scenario to imagine is that you are giving the presentation in a business setting to a group of engineers with whom you work daily, but the presentation you are giving was prepared in 2 minutes. You would be fairly informal in your speaking, but you would not shift into street slang. Certainly, the level of informality is a judgment call, but, when in doubt, err on the side of formality, particularly in speech patterns. In many ways, informal presentations are easier because the audience's expectations are lower than those for a formal presentation since they know you have not had much time to prepare. Nonetheless, you must be careful not to let your presentation level drop below the audience's expectations, and since you cannot know in advance what these expectations are, you should give a presentation as close to formal as possible.

8.6   Examples

The following examples are adapted from an actual presentation. Selected slides from this presentation were modified to show examples of good and poor slides.

Example 1. Figure 8.1 presents good and poor alternatives for a title slide. Figures 8.1a and 8.1b are good title slides. Figure 8.1b shows the authors' affiliation. Many people believe that showing the authors' affiliation is important, although as with all slides, since you are presenting the material orally, you can give the affiliation orally. Figure 8.1c shows a slide with too many words and too much information. The title is too long, and the positions and degrees of the authors are probably unnecessary. Again, some people would disagree and argue that titles and job positions are too important to omit from the slide. We believe the slides should show only essential information. You, as the presenter, should expand on the slide information as you see fit. *Remember:* This presentation is oral.

Example 2. Figure 8.2 shows good and poor alternatives for a list of ideas, in this case guidelines for analysis. Figure 8.2a shows a good slide. The total number of lines on the slide, including the title, is 8. Each idea, in this case a guideline, is a single short sentence. Each idea is explained orally during the presentation. Figure 8.2b shows a slide with too much information. The total number of lines, including the title, is 15. Each idea is explained in too much depth, with a single long sentence or two sentences. A slide with this many lines and this much information is generally not appropriate. One instance when a busy slide like this might be appropriate is when you are handing out copies of your slides and want the audience members to have all the information you are saying without their needing to take notes. The example shown in Figure 8.2 is from a presentation on a paper. The audience members can read the paper to remind themselves of the details.

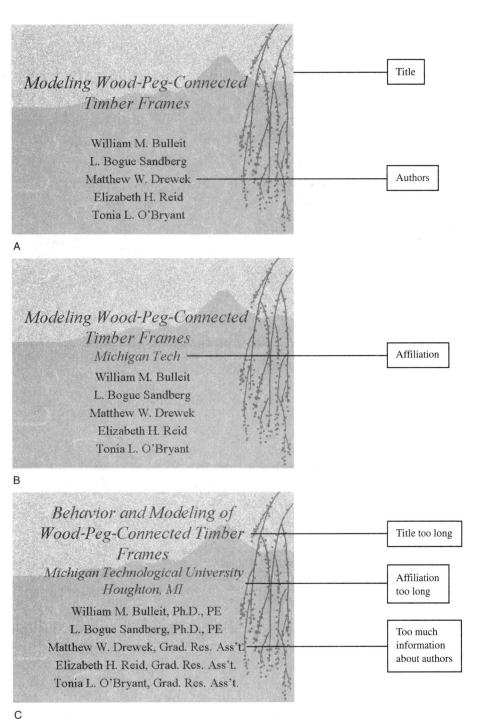

Figure 8.1. Title slides: (A) and (B) good slides; (C) poor slide

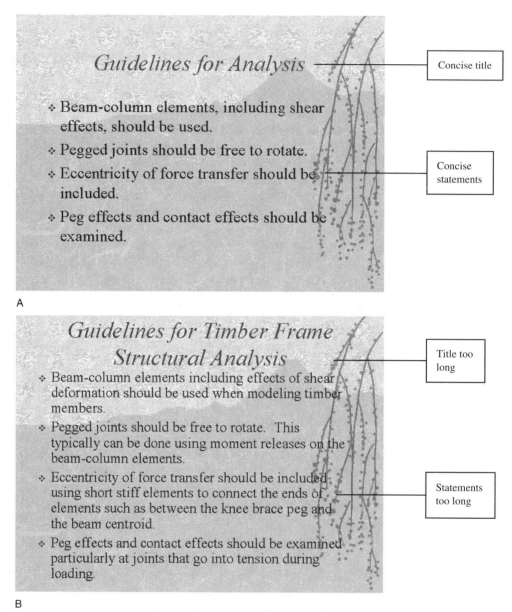

Figure 8.2. Slide with list of ideas: (A) good slide and (B) poor slide

Example 3. Figure 8.3 shows good and poor alternatives for a graph. Figure 8.3a shows an example of a graph with a reasonable amount of information, not too much to overload the viewers but enough to remind you of what you need to say. In contrast, Figure 8.3b has too much information. Viewers will be reading the slide rather than listening to your explanation of it.

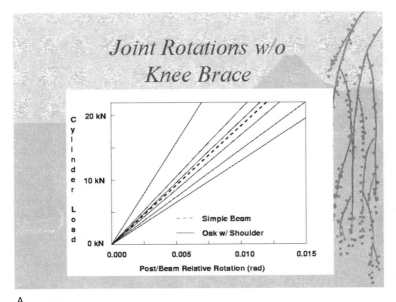

A

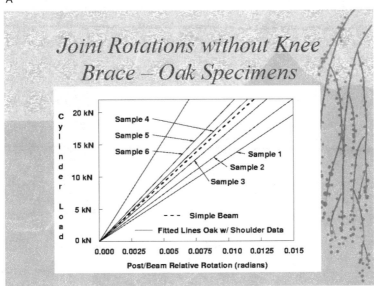

B

Figure 8.3. Slides with graphs: (A) good graph and (B) poor graph

Example 4. Figure 8.4 shows good and poor alternatives for a table. Figure 8.4a shows a table slide with an appropriate amount of information. The table has only four rows and four columns. Figure 8.4b shows a table with too many rows and columns. The table in Figure 8.4a was taken from the information shown in Figure 8.4b. Figure 8.4b might be appropriate for a table in a paper, when you want

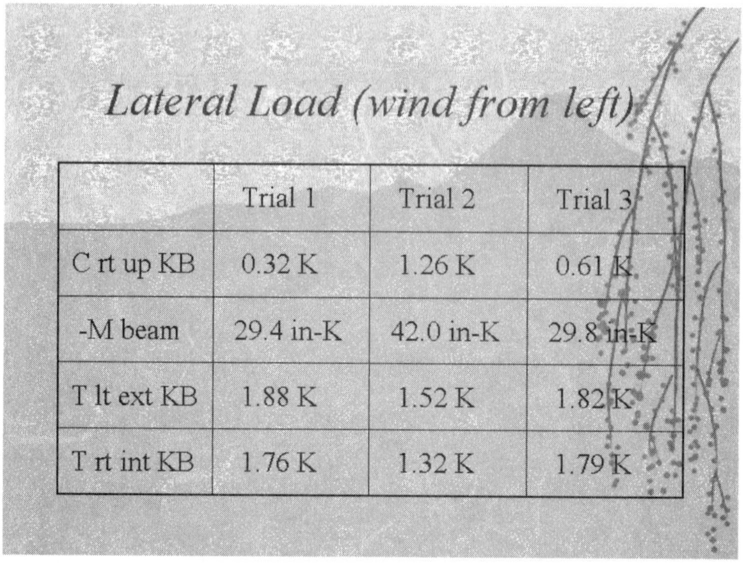

Lateral Load (wind from left)

	Trial 1	Trial 2	Trial 3
C rt up KB	0.32 K	1.26 K	0.61 K
-M beam	29.4 in-K	42.0 in-K	29.8 in-K
T lt ext KB	1.88 K	1.52 K	1.82 K
T rt int KB	1.76 K	1.32 K	1.79 K

A

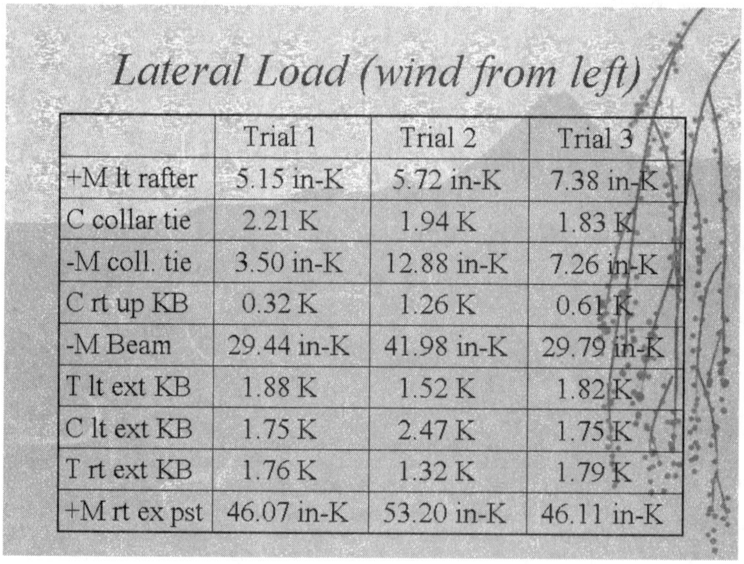

Lateral Load (wind from left)

	Trial 1	Trial 2	Trial 3
+M lt rafter	5.15 in-K	5.72 in-K	7.38 in-K
C collar tie	2.21 K	1.94 K	1.83 K
-M coll. tie	3.50 in-K	12.88 in-K	7.26 in-K
C rt up KB	0.32 K	1.26 K	0.61 K
-M Beam	29.44 in-K	41.98 in-K	29.79 in-K
T lt ext KB	1.88 K	1.52 K	1.82 K
C lt ext KB	1.75 K	2.47 K	1.75 K
T rt ext KB	1.76 K	1.32 K	1.79 K
+M rt ex pst	46.07 in-K	53.20 in-K	46.11 in-K

B

Figure 8.4. Slides with tables: (A) good table and (B) poor table

to show the reader all the information, but it is unacceptable for a presentation, when you should use only the information you are speaking on. If you are not going to talk about something, it should not be on a slide. Note, in Figure 8.4, that abbreviations are used in the table (e.g., C rt up KB, which stands for "compression in the right upper knee brace"). Abbreviations like this are acceptable in a presentation, but you must explain them when you discuss the slide.

8.7 Exercises

1. Prepare a checklist that can be used each time you must prepare an oral presentation. The list should include such items as the following: Does any slide contain too much information? Did I practice the presentation? Are all the slides necessary?

2. Select a journal article on a topic in your field and create the visuals for an oral presentation on it. Tailor your presentation to an audience of tenth-grade students.

3. Complete Exercise 2 tailored to an audience of your peers. Then give the presentation to a group of your peers and have them evaluate you.

4. Ideally, the presentation prepared in Exercise 3 should be given to a group of your peers who previously read the journal article. Make changes to your presentation materials as necessary. Ask the other group members to read your chosen article, then make the presentation to them and have them evaluate you.

5. Referring to Exercise 3 in Chapter 3 (ethics case studies), prepare a group presentation on the case you chose. Make sure you report the group's conclusions as part of your presentation.

6. Attend a university lecture or seminar. In a memo to your instructor, write a brief evaluation of the effectiveness of the presentation and compare and contrast the speaker's presentation style to that outlined in this chapter.

7. Find an audio speech on the Internet about a topic of interest to you. Create presentation visuals that could be used to augment the speech. Meet with a partner to view each other's work. Turn in to your instructor a memo evaluating your partner's presentation visuals. (Your partner should do the same.) Attach your set of visuals to your memo. Make sure you cite the location of the speech so that your instructor can listen to it if necessary.

8.8 Additional Resources

D'Arcy, J. 1998. *Technically speaking—A guide for communicating complex information.* Columbus, OH: Battelle Press.

Kenny, P. 1982. *A handbook of public speaking for scientists and engineers.* Bristol, UK: Adam Hilger.

Mablekos, C. M. 1991. *Presentations that work.* New York: IEEE Press.

Visual Communication

Visual communication is often an essential component of technical communication, and few technical documents or presentations can be considered complete without some graphical elements. In many cases, the old saying "A picture is worth a thousand words" is true. Imagine how many words would be required to describe a painting such as Leonardo da Vinci's *Mona Lisa* compared with the understanding instantly achieved by merely looking at the picture. Since one tenet of technical communication is that writing should be clear and *concise,* if using a visual aid means significantly reducing the volume of the required writing, the visual aid should be included if feasible. Another reason visual aids should be used in technical communication is that, by training, engineers are typically graphically oriented. Engineers often communicate informally with one another through hand-drawn sketches. These sketches are often drawn on the backs of envelopes or on paper napkins during lunch. Because engineers are so graphically oriented, judicious use of visual aids will allow your meaning to be understood better by other engineers. In essence, visual aids are a significant part of effective technical communication and most documents will be incomplete without at least one. In this chapter, we describe the various types of visual communication you may want to include in your technical documents and the accepted manner of such inclusion. As discussed in Chapter 8, visual aids are also vital to most oral communication.

Table 9.1. Tools for Graphical Communication

Tool	Tasks
Hand-drawn sketches	Hand-drawn sketches are important tools in technical communication. They are often faster and easier to create than drawings made with computer tools. In most cases, you should not rely on hand-drawn sketches for professional documents; however, for internal memos or letters, attaching a hand-drawn sketch is often one of the most effective forms of graphical communication.
Drawing tool embedded in a word processor or presentation software	The internal drawing tool found in most word-processing or presentation software is usually limited in capabilities. You can create simple graphics and basic shapes, such as rectangles, circles, and lines, with these tools, but they are not useful for creating complex drawings. In addition, most of these software packages have embedded clip art files that can be used to include graphical images in a document or a presentation. Clip art usually consists of images commonly used across a broad spectrum and will probably not be of particular usefulness in technical communication but may be of help in communicating with lay audiences.
Chart tool embedded in a word processor or presentation software	Many word-processing or presentation software packages have the ability to create simple charts and graphs within your documents. The charts and graphs can take many forms and be customized in multiple ways. If your chart or graph displays relatively large data sets, you may want to work in a spreadsheet (see next entry) to create the graph and then paste it into your document.
Spreadsheet	Spreadsheets are useful for creating graphs and charts that display large quantities of data. Charts and graphs can sometimes be pasted directly into your document, but you may need to write them to a separate file first and then import them into your document.
2-D computer-aided design (CAD) package	2-D CAD packages are useful for creating complex drawings when precision and exact dimensions are required. You will probably have to write your drawing to an external file and then import it into a document when working with this type of tool.

Table 9.1. (*continued*)

Tool	Tasks
3-D solid modeling package	3-D solid modeling packages are useful for creating 3-D or pictorial views of objects. You can use 3-D modeling packages to create a view of your object from any vantage point and can display objects with shaded surfaces or with lines at edges. Surface effects and different colors can be added to the shaded surfaces as desired. Once again, you will probably have to write the graphical images created in a 3-D modeling package to external files to import them into your document.
Scanners	Scanners are used to create electronic files from graphics for which you have only a hard copy. Scanners are useful for putting photographs or older, hand-drawn drawings into your document. The resolution of the scanned image (and therefore its appearance in your document) will not be as "clean" as it would be if the drawing file were used to create the image file, but sometimes you will have no alternative.

9.1   Tool Selection for Graphics

Many tools are available for creating graphical images for your reports, presentations, or other documents. Such tools have a wide range of capabilities and usefulness. Table 9.1 lists some of the more common tools available for creating graphics and the tasks for which these tools are most useful.

9.2   Representation of Data

Engineers routinely gather large quantities of data that they must then portray in technical documents. Two basic mechanisms are used to represent data within a document: tables and graphs (or charts). Several types of graphs and charts can be used for effective visual communication, and the type of graph or chart you select will usually depend on the type of data you want to portray. In this section, we describe common types of graphs and charts you may encounter, as well as general information about when and where each should be used in technical communication.

9.2.1 Common Features of Graphs and Charts

Effective graphs and charts share many features, regardless of the type of graphic chosen. In this section, we describe each feature and give general rules to be followed when you are creating graphs and charts. Exceptions to these rules are noted in the discussion of mechanics for each type of graph or chart. Five common features of graphs and charts are as follows:

1. Each axis on the chart or graph should be clearly labeled, and units should be included when appropriate. The horizontal axis of the chart should be labeled so that it is readable from the bottom of the upright page; the vertical axis label should be oriented so that it is readable from the right side of the upright page when the page is rotated 90 degrees clockwise. Numerical data values for both the horizontal and the vertical axes should be readable from the bottom of the upright page. Axes values should extend slightly beyond the maximum and minimum data values.

2. The graph should have a descriptive title across the top. This title should not be merely a repeat of the axis labels. For example, "Stress vs. Strain" is not an acceptable title for a graph. (An acceptable title would be "Tension Test Stress-Strain Results for A36 Steel.")

3. If more than one data set is represented on the graph or chart, you should include a legend that allows the person viewing the graph to understand which curve, bar, or wedge corresponds to each data set. The legend should be located so that it does not obscure the data presented on the graph. Legends are typically shown on the right side of the graph or across the top, just under the title. You should use the same hatching, color, line type, or data point marker in the legend that was used to create the graph.

4. In most cases, you should show the data points on the graph. Doing so will help the person viewing the graph discern values from the graph more accurately. If, however, including the data point symbols results in a cluttered graph that is difficult to understand, these symbols should be omitted. Clarity is always of primary importance.

5. Several software tools are available to help you create professional-looking charts and graphs. Most software tools create charts and graphs with a different color for each data set. If you do not have capabilities for or access to a color printer or photocopier, make sure you use different hatchings or gray scales if your graph contains more than one data set, since differences in color may not be obvious on the printed black-and-white document.

9.2.2 Pie Charts

Function. *Pie charts* may be used whenever you want to show the relative size of categories of the data you are representing, since each category on a pie chart represents a percentage of the whole. Pie charts are typically used to present nontechnical information (such as budgets) or to present material to nontechnical audiences.

Form. Pie charts are constructed as a circle divided into several wedge-shaped pieces, hence the name. Each data category is represented by one wedge on the pie chart, and the angle at the apex of the wedge is proportional to the relative size of the category. Since interior angles of the individual wedges must total 360 degrees, the angle of each wedge represents its percentage of this total angle. For example, for a data category that represents 25 percent of the total, the angle at the apex of this wedge would be 25 percent of 360 degrees, or 90 degrees. With modern-day software packages, pie charts can be created quickly and easily, can be color coded, and can include 3-D effects.

Mechanics. Three general rules should be followed when you are creating pie charts. First, have no more than six or seven individual pieces in the pie. Second, include the largest piece of the pie starting at the 12 o'clock position; include remaining pie pieces in descending order, according to size, clockwise relative to the largest piece. Third, include labels and data on the chart if possible. However, if you are using multiple pie charts to show differences between two or more sets of data, create them so that the same data set is always portrayed with the same shading or color and in approximately the same orientation (which may mean that on one of the pie charts the largest portion does not necessarily start at the 12 o'clock position). Since a pie chart has no axes per se, general rules about axis labels do not apply. Data portrayed on the chart can be entered either as percentages or as absolute numbers. If the data are included as absolute numbers, the size of each wedge should still be proportional to the relative size of the category with regard to the size of the whole.

If a color printer or photocopier is not available for producing your document, use hatching or shades of gray to distinguish among pie pieces. When using shades of gray to represent your pie pieces, make sure you select shades that are distinctly different. This precaution is especially necessary if you will be making multiple copies of your report since photocopiers often slightly darken shades of gray. Sometimes you may want to shade only one piece of the pie to give it special emphasis within the chart.

Examples
Figure 9.1 shows examples of pie charts both before and after they were edited to improve their appearance.

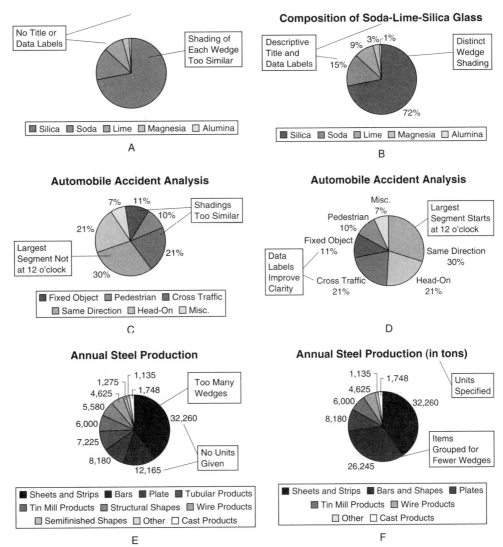

Figure 9.1. (A), (C), and (E) Poorly constructed pie charts. (B), (D), and (F) Pie charts improved from those in Figures 9.1A, 9.1C, and 9.1E, respectively.

9.2.3 Bar Charts

Function. *Bar charts* are commonly found in technical and nontechnical documentation. In fact, you have probably seen bar charts in newspapers and magazines as well as in technical textbooks and reports. Like pie charts, bar charts are well understood by the general public; however, bar charts are more useful for technical documentation because information and data can be displayed on them more efficiently than with pie charts.

Bar charts are useful for showing types of data that occur at discrete intervals or points in time rather than data that occur over a continuum. For example, as an environmental engineer, you may want to plot emissions from a smokestack during a 7-year period. Although time is a continuous function, this type of data could be displayed as a bar chart with individual bars representing emissions in year 1, year 2, year 3, and so forth. Creating a bar chart to represent these data would show not only whether annual emissions were decreasing during the given period, but also the numerical value for emissions in a given year.

Another use for bar charts is to show comparisons between two data sets. If you wanted to show the emissions of two smokestacks at a given plant, you might put the data on a bar chart on which data from one smokestack would be located adjacent to the data from the other smokestack. In this way, differences in emissions could be clearly seen by judging the relative heights of the bars. By viewing the chart, you could determine, for instance, that during a specific period, emissions were decreasing for one smokestack but holding constant for another smokestack.

Form. The data bars can be included on the bar chart either vertically or horizontally, although vertical bars are more commonly used. Bars can also be stacked to show cumulative data. For example, suppose you were a manufacturing engineer and wanted to show failure rates for parts created by various machines in the factory. In this case, failure is defined when dimensions on a given part are not within the allowable tolerance limits. For a given machine, you might have three gradations of failure: dimension variations of ±0.05–0.1 mm, ±0.1–0.25 mm, and more than ±0.25 mm. Bars showing failures in each category could be stacked to show the total number of failures per machine. In this way, total failures and failures in each category could be shown on a single graph rather than two or more graphs.

Mechanics. The width of each bar on the chart is not critical, but it should look proportionally correct on the overall chart. The width of all bars on the chart should be identical: if one were thicker than others, special emphasis would be implied. If several bars are used, each should be relatively thin; if only a few bars are used, each should be slightly thicker. Gaps slightly smaller than the width of the bars should be left between bars so that each data point is distinct.

If you want to show a comparison between two data sets, put the comparison bars next to each other with a space between them. You can show comparisons between more than two sets of data; however, you should ensure that the chart does not look too cluttered. Clarity of presentation is of primary concern.

If your data show points across time, make sure the years or months are included sequentially across the axis of the chart. If your bar chart does not show sequential data, you can arrange the bars in any order you choose; however, usually ascending or descending order is preferred.

Examples

Figure 9.2 shows examples of bar charts both before and after they were edited to improve their appearance.

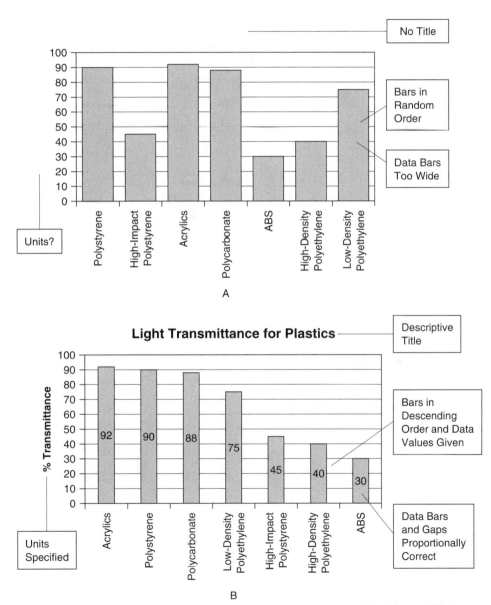

Figure 9.2. (A), (C), and (E) Poorly constructed bar charts. (B), (D), and (F) Bar charts improved from those in Figures 9.2A, 9.2C, and 9.2E, respectively.

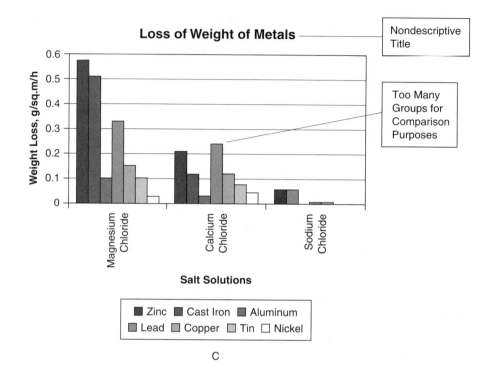

C

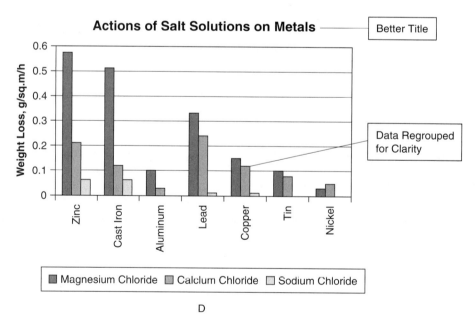

D

Figure 9.2. (A), (C), and (E) Poorly constructed bar charts. (B), (D), and (F) Bar charts improved from those in Figures 9.2A, 9.2C, and 9.2E, respectively. (*continued*)

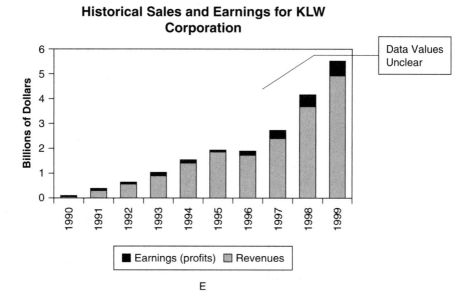

E

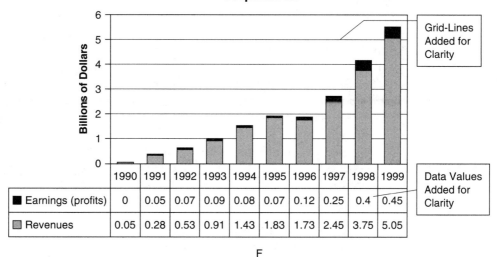

F

Figure 9.2. (A), (C), and (E) Poorly constructed bar charts. (B), (D), and (F) Bar charts improved from those in Figures 9.2A, 9.2C, and 9.2E, respectively. (*continued*)

9.2.4 **Histograms**

Function. A *histogram* is a special type of bar chart used to show the number of occurrences of given data values over the range of values. In other words, histograms show the distribution of data over a given range. Histograms are particularly useful for showing how the data were distributed within the data set in ways that statistical values such as an average cannot. For example, for relatively few data points, a single value could significantly alter the average value; however, a histogram would show that most data points were clustered around a given value and only one or two were at the extreme limits (commonly referred to as *outliers*). You are probably familiar with one common use of a histogram: when teachers show grade distributions for tests and quizzes. For example, a teacher may report that 5 students received As, 12 received Bs, 35 received Cs, 15 received Ds, and 4 received Fs on a given exam, and these results would be graphed as a histogram.

Form. Histograms are almost always shown as vertical rather than horizontal bars, and no gaps are used between individual bars. Data values along the horizontal axis are shown in ascending or descending (usually ascending) order, and the vertical height of a bar represents the number of occurrences of a specific data value. The data value along the horizontal axis can be a distinct value or a range of values. For example, in the case of reporting grade distributions for a given test, a data bar may be labeled either "A"—for a distinct, individual grade corresponding to a given data bar—or "90–100." Either way, the resulting data bar represents the number of students whose score fell within the A range. Data are typically presented as either the percentage of the whole or as absolute numbers. For example, in the case of student grade distributions, the percentage of students who earned each grade might be presented (i.e., 12 percent in the A range, 17 percent in the B range). Alternatively, the data could be presented as the number of students who earned each grade (i.e., 15 students received an A, 23 received a B). Either way, the total sample size should be included somewhere on the graph (sample size is usually indicated as $n = \#$). The sample size should be included because, statistically, a histogram showing data for a relatively large sample size is usually more meaningful than it would be for a small sample size.

Mechanics. The mechanics of creating a histogram are the same as those used to create a standard bar chart. One exception is that the *y*-axis label is typically "Number of _____" or "Percentage of_____," depending on the data set you are representing, and units are often not supplied. For example, in the case of exam scores, the title of the graph might be "Grade Distribution for Exam 1," the horizontal axis label might be "Grades" (with data values of A, B, C, D, F), and the vertical axis label might be "Number of Students" (with data values corresponding to the number of occurrences at each point along the horizontal axis).

Examples

Figure 9.3 shows two examples of histograms: one that portrays the data set as absolute numbers and one that portrays the data set as percentages. Note that in Figure 9.3, the data set did not consist of items that could be counted, such as the number of students in a given category. The weight of an aggregate sample remaining on a given sieve was portrayed (Figure 9.3a), then percentages were computed on the basis of the total weight of the sample (Figure 9.3b).

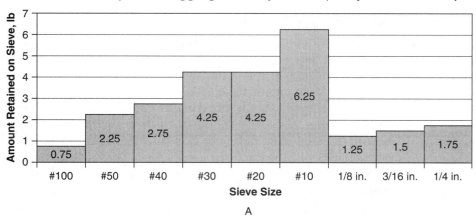

A

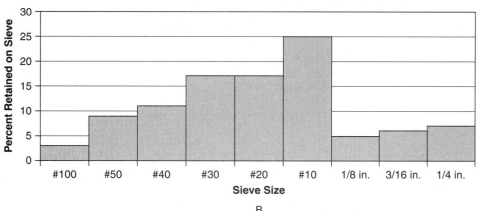

B

Figure 9.3. Histograms showing data as (A) absolute numbers and (B) a percentage of the total

9.2.5 Line Graphs

Function. *Line graphs* are used to show mathematical functions that are either linear or curved. Line graphs are one of the most common visual aids used to display data in technical documents. These graphs show data that occur over a continuous range rather than at discrete values. Line graphs can be used to show the trends (increasing or decreasing) and the rates of change (linear or exponential) of the data being presented. Actual data points are typically shown on the graph, as well as the idealized curve that best represents the data, so line graphs can also be used to show how well your experimental data fit the idealized conditions. Line graphs are typically used when you want to show several data sets on one graph, which allows comparisons between data sets and results.

Form. Curves and lines should be shown on the graph as continuous curves, and data points should be created as some type of symbol such as a small circle or an *X*. Different data sets typically have either different point styles or different curve styles. For example, one data set might have small circles as data points and another data set might have small triangles as data points. In other situations, you might choose to show one curve as a solid line and another curve as a dashed line. In either case, you should include a legend indicating which data sets belong to which data points or curves so that readers can readily determine the characteristics of the data being presented. The values of individual data points on the graph are not usually shown on the graph since generally a number of points are being presented. (Including all data point values would make the graph cluttered and difficult to read.) If you believe readers need to know the values of the data points represented on the graph, the data can be included in a table within the document. (Tables are described subsequently.)

Mechanics. Most word-processing, graphing, or spreadsheet software packages can be used to create line graphs easily; however, by default, they will usually "connect the dots" between data points. In almost all cases, dot-to-dot graphs are unacceptable in technical documents. The reason is that a graph in technical documentation represents data that occur over a continuous range. The data points were collected at discrete values within this range. Connecting the dots in this way would give a false impression that the data "jump" from point to point rather than occurring as a continuous function. (The exception to this restriction is for graphs that involve money or costs—these quantities typically do not occur over a continuous range but at discrete points.)

To create a line graph *correctly* in most software packages, you will have to turn off the curve that it automatically draws, leaving just the data points intact. You can then usually fit a line or a higher-order curve through the data points using an internal software function. One common approach is to use a *spline* function. The advantage of this approach is that you do not need to know what type of function fits the data. The disadvantage is that the spline function may produce a curve that varies wildly around and through the data points. If a "wild" curve occurs, you will need to use something other than a spline because this type of curve is no more realistic than the curve that results from connecting the dots. Alternatively, you may have to plot the data points as a scatter plot using the software package and fit lines or curves through the data on the graph. If your software package cannot fit a curve through your data points, you may have to determine the equation of the curve with a calculator, calculate several points from this equation, and then plot the result on your graph, this time turning off the data points and leaving the curve turned on.

In many cases, you should include the equation of the function shown on the graph if possible. You should do so especially if you have fit a line to the data since the slope and intercept of a linear function are usually of particular interest to anyone viewing the graph. If you use a spline, you generally should not show the equation: just state in the body of the text that a spline function was used.

Examples

Figure 9.4 shows several examples of line graphs both before and after they were edited to improve their appearance.

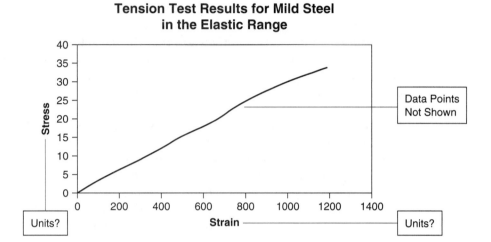

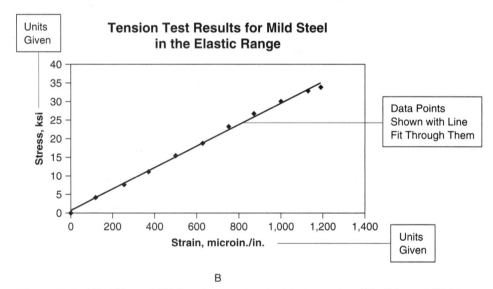

Figure 9.4. (A), (C), and (E) Poorly constructed line graphs. (B), (D), and (F) Line graphs improved from those in Figures 9.4A, 9.4C, and 9.4E, respectively.

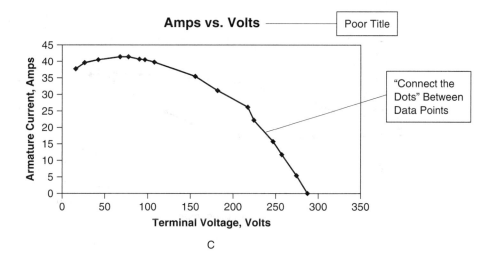

C

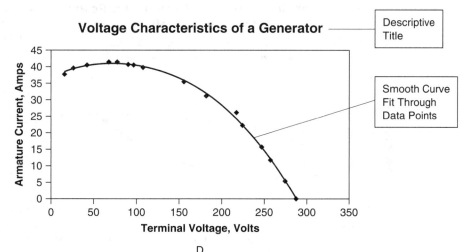

D

Figure 9.4. (A), (C), and (E) Poorly constructed line graphs. (B), (D), and (F) Line graphs improved from those in Figures 9.4A, 9.4C, and 9.4E, respectively. (*continued*)

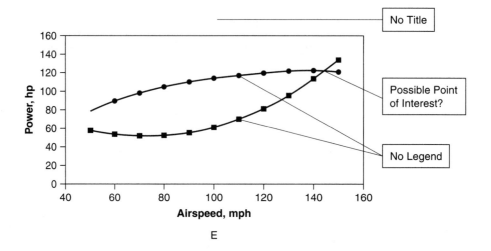

E

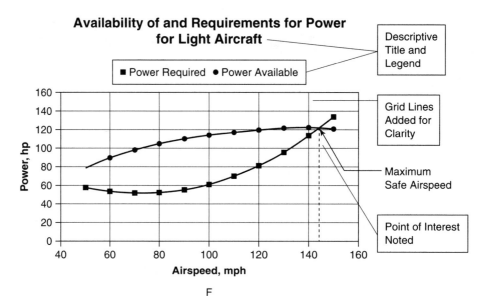

F

Figure 9.4. (A), (C), and (E) Poorly constructed line graphs. (B), (D), and (F) Line graphs improved from those in Figures 9.4A, 9.4C, and 9.4E, respectively. (*continued*)

9.2.6 Scatter Plots

Function. *Scatter plots* are a special type of line graph commonly used to show relatively large amounts of data with a high degree of variability in the data. Scatter plots are typically not conducive to curve fitting, and the primary purpose of the graph is to show the extent of the data variability and any trends in the data. The resulting graph often looks as if someone just splattered points around on the grid. Sometimes, however, the points will show a trend in the data (e.g., upward or downward). Unlike other types of graphs discussed previously in this chapter, scatter plots rarely show more than one data set, so comparisons between data sets on a single graph are not possible.

Form. On scatter plots, data points are shown by a symbol, but curves are not generally included. An exception to this rule is when a line is fit through the points and statistical details such as the coefficient of determination (R^2) are included. A detailed description of the coefficient of determination is beyond the scope of this text (you will likely learn about it in your statistics course); however, a brief description follows. The coefficient of determination indicates how well the data fit the line: a coefficient of determination of 1 means the data fit perfectly, a coefficient of determination of 0 means no line fits. In general, coefficients of determination of 0.6 and greater are considered adequate for statistical purposes. As you create scatter plots, most commercial graphing software will automatically determine the coefficient of determination.

Mechanics. If you are using a software package to create a scatter plot, be sure to turn off the curve that is automatically generated, and just show the data points. Be sure the symbols you use to represent individual data points are large enough so that they are clearly visible yet small enough so that they do not obscure one another. Do not include curves unless you want to show how well the data correlate to a straight-line fit. If you do include the trend line, be sure to also include its slope, the *y* intercept, and the coefficient of determination for the fit of the data.

Examples

Figure 9.5 shows examples of the type of scatter plots that might be found in technical documents.

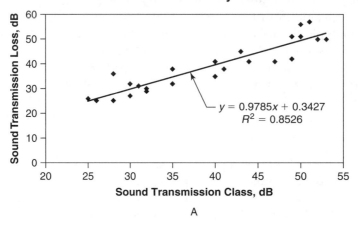

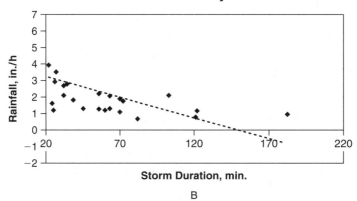

Figure 9.5. Scatter plot with (A) statistical information and best-fit line included and (B) best-fit line included

9.2.7 Tables

Function. *Tables* are perhaps one of the most common forms of portraying data in technical documentation. Tables can show either numerical data or nonnumerical data and are therefore more flexible than graphs, which can show only numerical data. Sometimes tables will accompany graphs within technical documentation and include the numerical values of all the data points used on the graph. In this way, the graph can give the audience a broad view of the data, with trends and rates of change clearly visible, while the accompanying table gives the audience the details about the data values. Tables can also be used to portray information that occurs over a matrix of values and are thus effective for comparing groups with respect to several parameters at one time. For example, if you were weighing the advantages of several computer systems you were considering purchasing, you could arrange the model numbers of the various systems along the columns of the table and the different features such as price, RAM, disk space, and so forth along the rows of the table. By viewing the matrix of information and weighing the pros and cons of each system, you could make a logical decision about the model you should purchase.

Another way to think about data representation is that 2-D graphs clearly show functions with one independent variable (i.e., y as a function of x), whereas tables can be used to show functions with several independent variables. In the example of comparing computers for purchase, *price* is a function of *features,* such as RAM and disk space; *model* or processor; and *manufacturer,* among other things.

Form. Tables comprise a number of rows and columns. You will set up your table according to the number of data points in the set. The intersection of a row with a column is called a *cell,* similar to the nomenclature used in spreadsheet software. Each row and each column of your table should be labeled so that other individuals know what data are included. In a document, tables should have titles as well, similar to the titles included with graphs and charts. In some cases, you may want to group several rows or columns with sublabels under a single umbrella label. You can then straddle or merge the cells of the main label, with the sublabels located either underneath or to the right of the main label.

Mechanics. All modern tools such as word processors and presentation software packages have the ability to include tables in your documentation. You can also format your tables to make them appear neat, well organized, and professional. Data typically look better if they are centered within the cells; however, sometimes you may want the data to be right or left justified. When tables include money amounts in the cells, you may want to right justify the numbers throughout a table column so that decimal points align.

You are free to treat the text within cells differently from one another, applying bold and italics for emphasis in individual cells, especially for column and row labels. You can also add special shading or borders to individual cells or to groups of cells to provide emphasis. When adding these special effects, however, you must not use too many, or a table will result that is difficult to understand. As for all forms of technical communication, clarity and conciseness should be your two guiding principles.

Tables are normally centered within the page or column of text in your document, but centering is not always a requirement. Sometimes, such as for journal publications, you will be given explicit instructions about the desired final appearance of your table.

Examples
Figure 9.6 shows several examples of tables both before and after they were edited to improve their appearance.

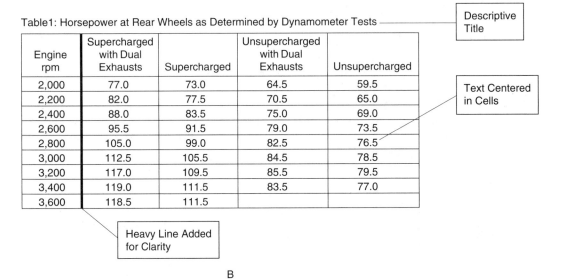

Figure 9.6. (A), (C), and (E) Poorly constructed tables. (B), (D), and (F) Tables improved from those in Figures 9.6A, 9.6C, and 9.6E, respectively.

Column Headings "Clunky"

Table 7. Dimensions of External-Tooth Lock Washers

Washer Number	Nominal Washer Size	Max. Inside Diameter	Min. Inside Diameter	Max. Outside Diameter	Min. Outside Diameter	Max. Thickness	Min. Thickness
3	0.099	0.109	0.102	0.235	0.220	0.015	0.012
4	0.112	0.123	0.115	0.260	0.245	0.019	0.015
5	0.125	0.136	0.129	0.285	0.270	0.019	0.014
6	0.138	0.150	0.141	0.320	0.305	0.022	0.016
8	0.164	0.176	0.168	0.381	0.365	0.023	0.018
10	0.190	0.204	0.195	0.410	0.395	0.025	0.020
12	0.216	0.231	0.221	0.475	0.460	0.028	0.023

Units?

C

Units Specified

Column Headings Grouped for Clarity

Table 7. Dimensions of External-Tooth Lock Washers (All dimensions in inches)

Washer Number	Nominal Size	Inside Diameter		Outside Diameter		Thickness	
		Maximum	Minimum	Maximum	Minimum	Maximum	Minimum
3	0.099	0.109	0.102	0.235	0.220	0.015	0.012
4	0.112	0.123	0.115	0.260	0.245	0.019	0.015
5	0.125	0.136	0.129	0.285	0.270	0.019	0.014
6	0.138	0.150	0.141	0.320	0.305	0.022	0.016
8	0.164	0.176	0.168	0.381	0.365	0.023	0.018
10	0.190	0.204	0.195	0.410	0.395	0.025	0.020
12	0.216	0.231	0.221	0.475	0.460	0.028	0.023

Thick Cell Borders Added for Readability

D

Figure 9.6. (A), (C), and (E) Poorly constructed tables. (B), (D), and (F) Tables improved from those in Figures 9.6A, 9.6C, and 9.6E, respectively. (*continued*)

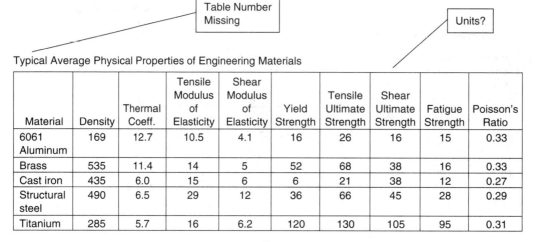

Typical Average Physical Properties of Engineering Materials

Material	Density	Thermal Coeff.	Tensile Modulus of Elasticity	Shear Modulus of Elasticity	Yield Strength	Tensile Ultimate Strength	Shear Ultimate Strength	Fatigue Strength	Poisson's Ratio
6061 Aluminum	169	12.7	10.5	4.1	16	26	16	15	0.33
Brass	535	11.4	14	5	52	68	38	16	0.33
Cast iron	435	6.0	15	6	6	21	38	12	0.27
Structural steel	490	6.5	29	12	36	66	45	28	0.29
Titanium	285	5.7	16	6.2	120	130	105	95	0.31

E

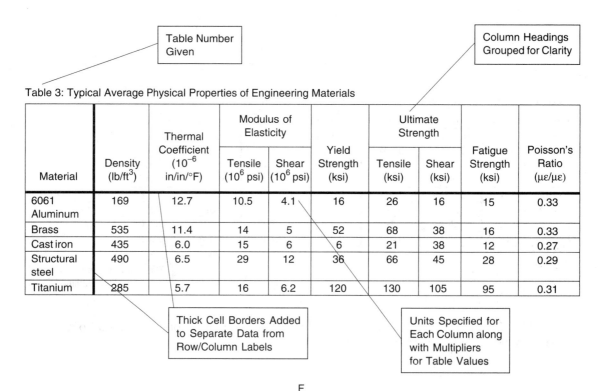

Table 3: Typical Average Physical Properties of Engineering Materials

| Material | Density (lb/ft^3) | Thermal Coefficient (10^{-6} in/in/°F) | Modulus of Elasticity | | Yield Strength (ksi) | Ultimate Strength | | Fatigue Strength (ksi) | Poisson's Ratio (µε/µε) |
			Tensile (10^6 psi)	Shear (10^6 psi)		Tensile (ksi)	Shear (ksi)		
6061 Aluminum	169	12.7	10.5	4.1	16	26	16	15	0.33
Brass	535	11.4	14	5	52	68	38	16	0.33
Cast iron	435	6.0	15	6	6	21	38	12	0.27
Structural steel	490	6.5	29	12	36	66	45	28	0.29
Titanium	285	5.7	16	6.2	120	130	105	95	0.31

F

Figure 9.6. (A), (C), and (E) Poorly constructed tables. (B), (D), and (F) Tables improved from those in Figures 9.6A, 9.6C, and 9.6E, respectively. (*continued*)

9.2.8 Three-Dimensional Charts

Function. Most modern word processors include the ability to create *3-D charts*. Three-dimensional charts are particularly useful if you are presenting data that are a function of more than one variable. Typically, the two independent variables are plotted along the two horizontal axes of the 3-D space occupied by the graph, and the dependent variable is plotted along the vertical axis.

Form. Virtually every type of 2-D graph described in the preceding sections of this chapter could also be created in a 3-D format. Three-dimensional pie charts, bar graphs, line graphs, and histograms are all possible. (The corollary to a 2-D line graph is a 3-D *surface graph*.) As for other types of graphs, you should include a title, axes labels (with units), and other relevant information on a 3-D graph so that it can stand alone within the document.

Mechanics. Just because 3-D charts are possible does not mean they should always be used. If the third dimension adds nothing to the clarity or comprehension of the data being presented, just use a 2-D chart. The same rules for the creation of 2-D graphs are applicable to the creation of 3-D plots. One additional consideration for 3-D graphs is to select an orientation so that the smaller values are "in front" and don't obscure the data being displayed toward the back.

Examples

Figure 9.7 shows some examples of 3-D graphs that might be useful in technical documentation.

Automobile Accident Analysis

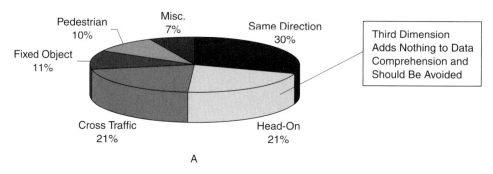

Pedestrian 10%

Misc. 7%

Same Direction 30%

Fixed Object 11%

Third Dimension Adds Nothing to Data Comprehension and Should Be Avoided

Cross Traffic 21%

Head-On 21%

A

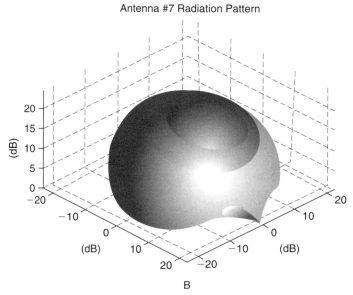

Antenna #7 Radiation Pattern

B

Figure 9.7. (A) Three-dimensional pie chart. (B) Three-dimensional mesh plot. (C) Poor example of a 3-D bar chart. (D) Bar chart improved from that in Figure 9.7C.

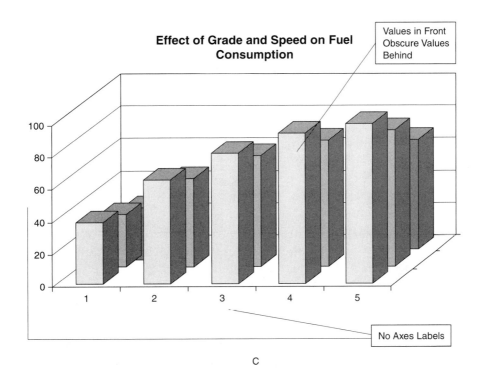

C

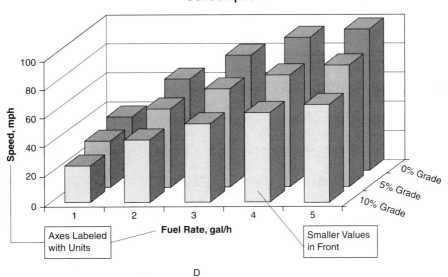

D

Figure 9.7. (A) Three-dimensional pie chart. (B) Three-dimensional mesh plot. (C) Poor example of a 3-D bar chart. (D) Bar chart improved from that in Figure 9.7C. (*continued*)

9.3 Charting Structure

A large part of any engineering position is devoted to managing projects that can vary in duration from a few weeks to several years. Throughout each project period, several reports will typically need to be generated about project planning and progress. In this type of documentation, graphical images are often needed to show structure or flow. In this case, *flow* can mean the flow of materials, the flow of information, or the flow of time; *structure* can mean a reporting structure or a decision-making structure. The types of graphical images used to portray these ideas are described in the following sections.

9.3.1 Flowcharts

Function. *Flowcharts* consist of boxes and arrows used to chart a process. Each box contains a step or an individual activity in the process, and arrows show the flow between steps. Sometimes an individual step in the process can have more than one possible outcome. This possibility is indicated on the flowchart by showing more than one arrow emanating from the given box.

Form. The boxes on a flowchart will usually have different shapes (rectangles, diamonds, ovals, circles, etc.) and will be grouped by shape to indicate similar types of activities. For many applications, most shapes are standardized according to context. For example, diamonds are used when more than one possible exit path exists (for a yes-no decision). For a flowchart you could create, you might use rectangles to represent fieldwork to be conducted and circles to represent activities conducted in the office to support the fieldwork (e.g., completing purchase requisitions). Each box on the flowchart should represent a single step in the process and have a two- to four-word descriptive title. Be careful not to include too many words on your flowchart. As for other types of technical communication, clarity is of primary importance and too many words will clutter your flowchart and make it difficult to understand.

Mechanics. Arrows on flowcharts are usually relatively thick, and 90-degree turns in the arrow path show changes in direction. The flowchart should be limited to a single page (or part of a page). Since a flowchart is meant to represent actions undertaken sequentially within a process, you should have the "flow" be either from the top of the page to the bottom or from the left-hand side of the page to the right. Having a process flow from bottom to top or from right to left would only confuse other people. The flowchart should show just the major activities of the process; details can be left to other parts of the document. You should use only a few shapes for your boxes on the flowchart: too many shapes will be distracting.

Examples

Figure 9.8 shows examples of the types of flowcharts found in technical documentation.

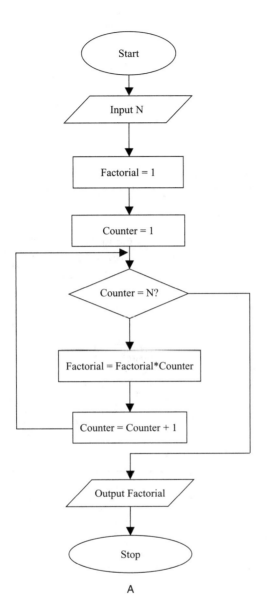

A

Figure 9.8. Flowcharts (A) for a computer program used to calculate a factorial, (B) for an electric car controller, and (C) illustrating a project completion cycle

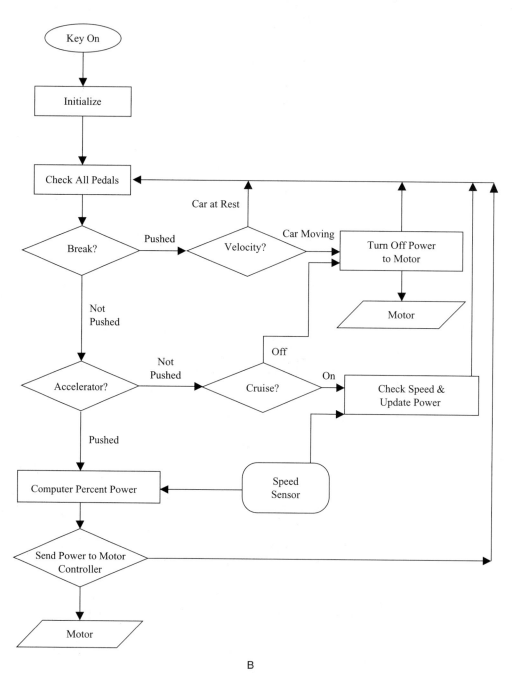

B

Figure 9.8. Flowcharts (A) for a computer program used to calculate a factorial, (B) for an electric car controller, and (C) illustrating a project completion cycle (*continued*)

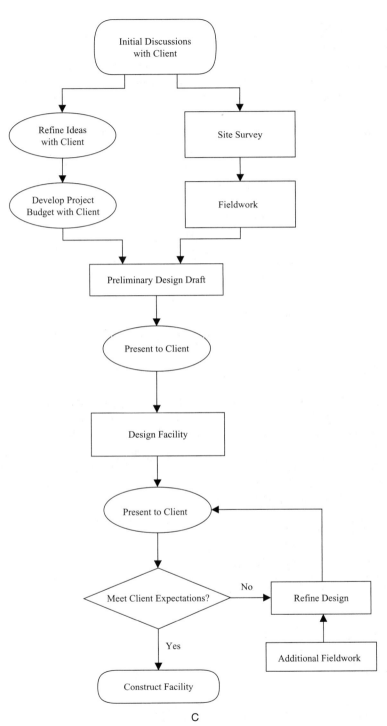

C

Figure 9.8. Flowcharts (A) for a computer program used to calculate a factorial, (B) for an electric car controller, and (C) illustrating a project completion cycle (*continued*)

9.3.2 Organizational Charts

Function. The most frequent use of an *organizational chart* is to show a reporting structure within a given company. In some cases, an organizational chart may be useful for showing a reporting structure for a large project on which several engineers and technicians are working.

Form. An organizational chart is set up similarly to a family tree. The person who is ultimately responsible for a company or a project is at the top of the chart; the persons who report directly to this top-level person are aligned on a second row, with lines connecting them to the top-level person. Persons who report to the managers found on the second tier are listed on a third row, and so on. Thus, the organizational chart "branches out" as you move from top to bottom within the organization. Tree structures can also be shown so that branching occurs from the bottom (root) to the top (leaves). This type of branching is often found on family trees and decision trees.

Mechanics. Titles are usually included in an organizational chart in place of individual names. For example, if Claire Hoffman serves as Affirmative Action Officer for a given company and reports directly to the president, her title (i.e., Affirmative Action Officer), rather than her name, would be listed on the organizational chart for the company. In this way, if she moves on to a different job within the company or leaves the company, the organizational chart does not need to be updated to reflect this change. In general, only rectangular boxes should be used to represent the various positions within the company. A variety of shapes will likely not serve any real purpose and will only confuse the reader. Positions should be aligned in tiers so that first-level managers all appear on the same row within the organizational chart. The boxes representing company positions should be connected by solid lines to show the reporting structure. Dashed lines can be used to show a reporting structure that is only advisory.

Examples
Figure 9.9 shows examples of organizational charts that might be used in technical documentation.

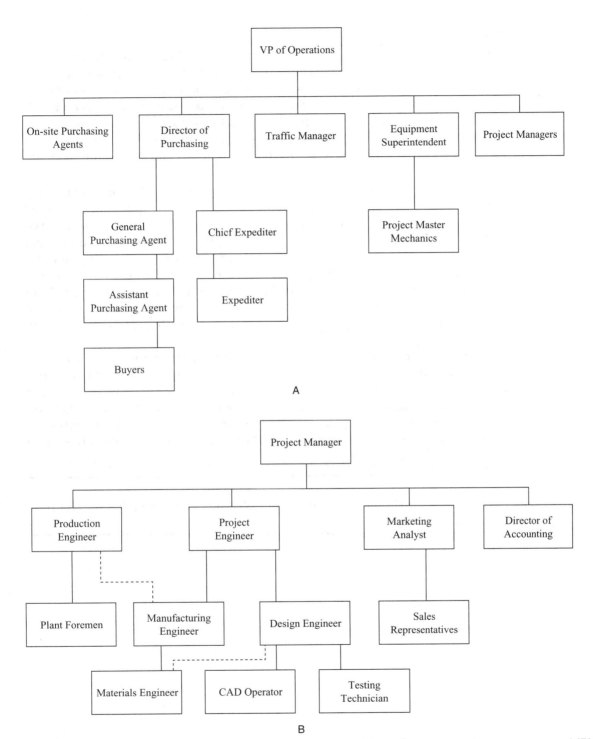

Figure 9.9. Organizational charts of (A) an operations division of a construction company and (B) a project management structure

9.3.3 Gantt Charts

Function. A *Gantt chart* is used to portray a project schedule graphically. Since most large-scale projects can be divided into several tasks that must be accomplished, the Gantt chart typically portrays the schedule by task or by milestone. A Gantt chart enables the reader to view, at a glance, significant milestones and tasks and their estimated time to completion. Gantt charts can be used in a proposal to show estimated times to completion for each task within the proposed work. Gantt charts can also be used in a progress report to show completed tasks and unfinished tasks.

Form. A Gantt chart is set up like an *x-y* graph, with project milestones along the *y* axis and time to completion along the *x* axis. The project milestones are listed sequentially along the vertical axis; initial tasks are listed at the top of the axis and final tasks at the bottom. Project time along the *x* axis can be listed in either absolute or relative terms. You should specify either starting and ending dates for each task along this axis or the number of days or weeks from the start date. The duration of each task is shown on the Gantt chart as a shaded bar. Realistically, in a large-scale project, a good deal of overlap occurs between ending one task and starting another, so this overlap must be reflected in the bars of the Gantt chart.

Mechanics. List only the major tasks or activities along the vertical axis of the Gantt chart. If necessary, for each major task you can create a separate Gantt chart that shows the details behind the task. Gantt charts can easily be created in most word-processing software packages through the use of the *Table* functions. Table cells can be shaded to create the horizontal bars of the chart, with tasks included in the leftmost table column and time included along either the top or the bottom row of the table.

Examples

Figure 9.10 shows several examples of Gantt charts suitable for technical documentation.

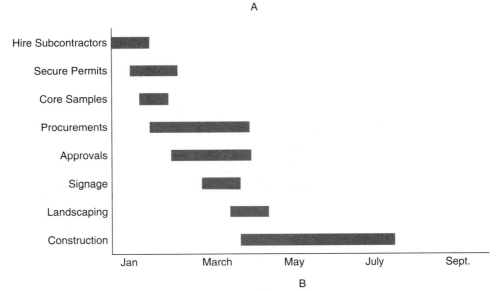

	Jan.	Feb.	Mar.	Apr.	May	June	July	Aug.	Sept.	Oct.	Nov.
Preliminary Design	▓	▓									
Design Revisions		▓	▓	▓							
Project Bids					▓						
Site Preparation					▓	▓					
Construction						▓	▓	▓	▓		
Finishing									▓	▓	
Move In											▓

A

B

Figure 9.10. Gantt charts (A) illustrating a home construction schedule, (B) suitable for a project proposal, and (C) suitable for progress reports

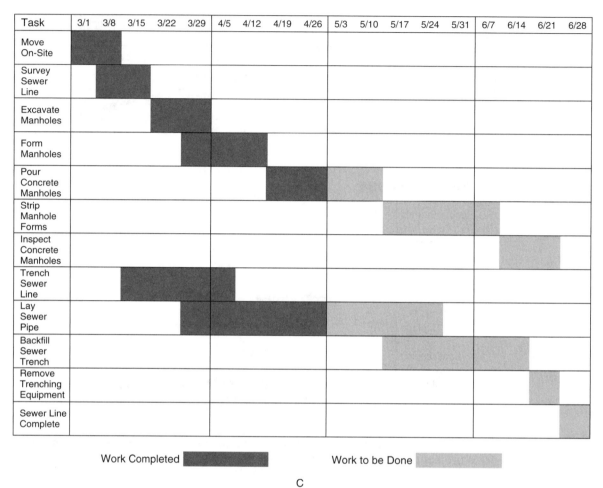

Task	3/1	3/8	3/15	3/22	3/29	4/5	4/12	4/19	4/26	5/3	5/10	5/17	5/24	5/31	6/7	6/14	6/21	6/28

Work Completed   Work to be Done

C

Figure 9.10. Gantt charts (A) illustrating a home construction schedule, (B) suitable for a project proposal, and (C) suitable for progress reports (*continued*)

9.4 Portrayal of Realism

Most of the types of graphical images discussed to this point in this chapter have been focused on abstractions: graphing functions in space or time, depicting the flow of time or material, showing data points within a continuum, and so forth. In some cases, however, you, as an engineer, will need to be able to portray an object, a system, an experimental setup, or some other feature you are describing realistically within the written or verbal portion of your technical documentation. In this section of the chapter, we discuss situations in which realism in graphics is required and the types of graphical images you may choose from to portray this realism. As shown in Figure 9.11, engineering drawings typically

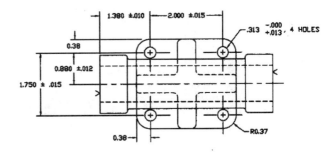

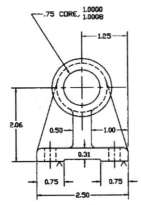

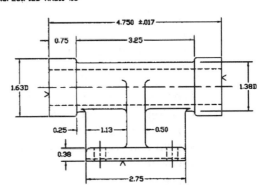

Figure 9.11. Engineering drawing (3-D Visualization for Engineering Graphics by Sorby/Manner/Baartmans, ©1998. Used by permission of Pearson Education, Inc., Upper Saddle River, NJ)

consist of several orthographic views, special views such as sections or details, and dimensions and other forms of annotation. Such drawings are a significant form of graphical technical communication. Creation of these technical drawings is a separate course of study for most engineers and is not covered in this text. A number of excellent engineering graphics texts on the market specifically cover the creation of and standards for engineering drawings. You should refer to these resources if you are specifically interested in this topic (see Section 9.8).

9.4.1 Schematic Drawings

Function. *Schematic drawings* are used to portray a complex system in a simplified manner. For example, Figure 9.12 shows a schematic drawing of the operation of a lathe. A lathe is a complicated piece of equipment with many moving parts; however, this schematic drawing shows the essence of how a lathe works without the clutter of unnecessary detail. In another case, a schematic drawing may be used to portray an electrical circuit, with symbols representing resistors, capacitors, and so forth. In this type of schematic drawing, the symbols represent real

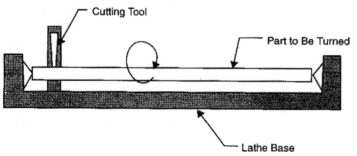

Figure 9.12. Schematic drawing of a lathe (3-D Visualization for Engineering Graphics by Sorby/Manner/Baartmans, ©1998. Used by permission of Pearson Education, Inc., Upper Saddle River, NJ)

objects but do not appear realistically on the drawing. Figure 9.13 shows some of the standard symbols used in electrical circuit diagrams.

Form. No specific form is used for a schematic drawing; that is, the appearance of the drawing will depend largely on the system you want to portray. In general, however, simplicity should be the rule. Show only the essential system components. If some system components are not needed to understand how the system operates, omit them from the schematic. In the case of electrical circuit diagrams, make sure all components are labeled with values given. Include the resistance of each resistor on

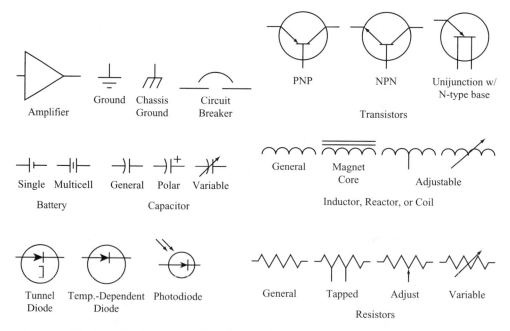

Figure 9.13. Standard symbols for electronic graphics

the diagram, the voltage of each battery, and the capacitance of each capacitor. For example, a resistor could be labeled "12 K ohms" or a battery "+12 V."

Mechanics. Most word-processing packages have embedded drawing tools that can be used to create simple schematic drawings. When feasible, include shading as appropriate to distinguish between parts on the schematic, and use color in your drawing only if you have the ability to reproduce a color document. Replace complex shapes in the system with simpler shapes on the schematic whenever possible. In the case of electrical circuit diagrams, you can often create a library of symbols, similar to those shown in Figure 9.13, that can be placed on the drawing where needed. Connecting lines can then be added between the symbols to result in the final drawing. The American National Standards Institute (ANSI; www.ansi.org) has developed standards to regulate engineering drawing practices. The standard that regulates the creation of circuit diagrams is ANSI Y14.15, and the symbols are regulated by ANSI Y32.2.

Examples
Figure 9.14 shows several examples of schematic drawings suitable for technical documentation.

9.4.2 Pictorial Drawings

Function. *Pictorial drawings* are used to portray an object or a system of objects in 3-D space on a 2-D piece of paper. Pictorial drawings are used when a realistic view of an object or a system is required for understanding. One common use of pictorials is for assembly drawings. Assembly drawings are used to show how all the pieces of a system fit together. For this reason, accurate, realistic pictorial views of the objects are required, thus schematic drawings are unsuitable for this type of graphical representation.

Form. You should be familiar with two basic types of pictorial drawings: isometric and oblique. Figure 9.15 shows isometric and oblique pictorial views of a simple cube. Isometric drawings show a 3-D part or system from a perspective defined as if you were looking down the diagonal of a cube with respect to the object. Isometric pictorials show all surfaces on the object in a distorted manner (e.g., squares will appear as parallelograms). Oblique pictorials are drawn so that one face of the object is parallel to your viewing plane and the third dimension is shown at an oblique angle from there. With oblique pictorials, one face of the object remains undistorted.

Mechanics. Realistic pictorial drawings are probably too complicated to draw with the tools available in most word-processing software packages. Most computer-aided drafting software packages can be used to create pictorial drawings. Alternatively, objects created with solid modeling software can be used to create pictorial drawings. Creation of pictorial drawings is beyond the scope of this text. Refer to

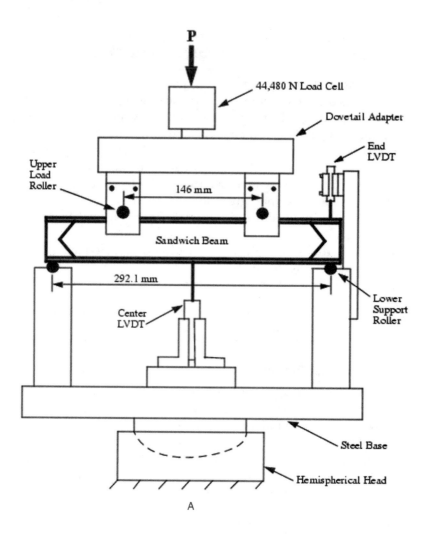

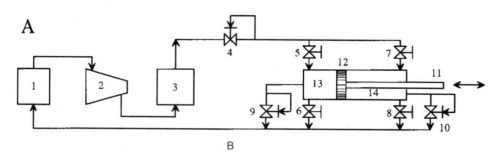

Figure 9.14. Schematic drawings of (A) a experimental test setup, (B) an actuator, and (C) electrical circuits

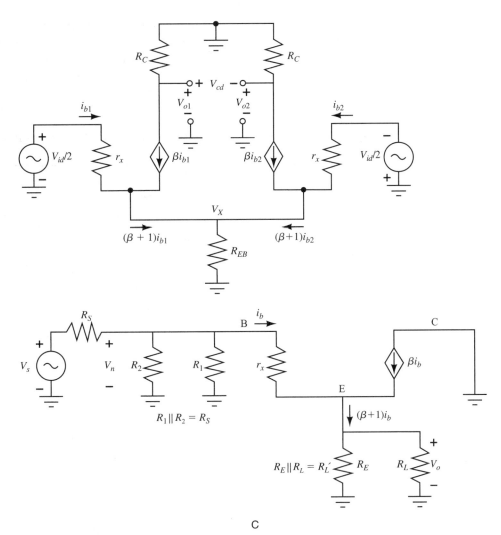

Figure 9.14. Schematic drawings of (A) a experimental test setup, (B) an actuator, and (C) electrical circuits (*continued*)

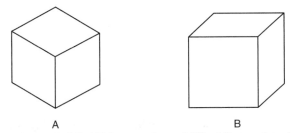

A   B

Figure 9.15. (A) Isometric and (B) oblique pictorial views of a cube (3-D Visualization for Engineering Graphics by Sorby/Manner/Baartmans, ©1998. Used by permission of Pearson Education, Inc., Upper Saddle River, NJ)

any one of the engineering graphics texts listed in the Additional Resources section of this chapter to help you create this type of drawing. Within the context of effective technical communication, you should make sure that your pictorial drawings in documents have parts that are labeled for clarity and that they show sufficient detail to represent the object or system accurately.

Examples

Figure 9.16 shows several examples of pictorial drawings suitable for technical documentation.

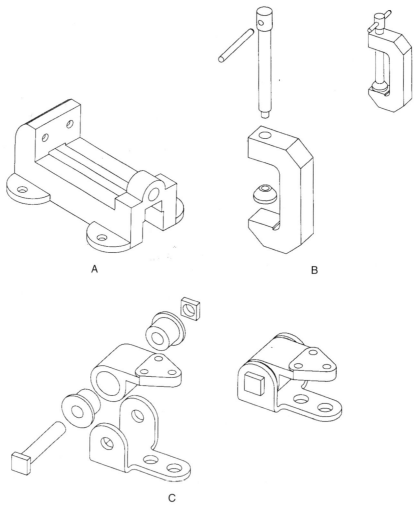

Figure 9.16. (A) Isometric pictorial of a mechanical vise. (B) Pictorial assembly drawing of a C-clamp. (C) Pictorial assembly drawing for technical documentation. (Solid Modeling with I-DEAS by Sorby, ©2003. Used by permission of Pearson Education, Inc., Upper Saddle River, NJ)

9.4.3 **Photographs**

Function. *Photographs* are frequently used in nontechnical documentation, such as brochures and marketing materials. For technical communication purposes, photographs should be used sparingly and only when absolute realism is required for clear communication. One of the most common uses for photographs is illustrating a particular experimental setup that might be difficult to describe with words or a pictorial drawing. Another use of photographs might be in a report that describes damage to structures or systems. In this case, photographs are taken of the damage for documentation purposes.

Form. Photographs used in technical documentation can be in color or in black-and-white; however, black-and-white is usually preferred. In most cases, avoid including people in photographs unless absolutely necessary. When photographing lab setups, you should clear clutter from the area and focus on only the specific features you want to illustrate. If you need a picture of a piece of equipment, set it on an empty table before photographing it. Reproduction of photographs in reports is not always optimal, since most photocopiers will darken and blur originals. For this reason, make sure your photographs are light, high contrast, and sharply focused.

Mechanics. Digital cameras are now relatively inexpensive and simple to operate, which makes including photographs in documents relatively easy. When capturing the digital images, be sure to strike a balance in setting the resolution. If your resolution is too low, your images will be fuzzy and indistinct; if your resolution is too high, your electronic file size will be large and you will have difficulty working with your document. If you do not have a digital camera available, you can always create photos with film and scan them to produce an electronic file. Once again, however, you should take care when setting the resolution for scanning so that you obtain crisp, clear images that don't bog down your computer system.

Examples
Figure 9.17 shows examples of photographs that could be used in technical documentation.

9.5 **Inclusion of Graphics in Documentation**

Graphics, in the form of figures and tables, are an essential part of effective technical communication. In this chapter, we presented several types of graphical images you can use throughout technical documentation to demonstrate ideas, objects, and data clearly. One key issue remains: *how* the graphical images are cited and included within the document. A poor practice is to merely "plop" the image into the body of a document. Setting the stage for the image is of

A   B

Figure 9.17. Photographs, suitable for inclusion in a technical document, showing (A) deterioration of an I beam end due to leaky bridge joints and (B) equipment used in experimental work

paramount importance so that a reader understands where the graphic fits within the context of the document and does not need to spend a lot of time determining exactly what the image means. This section contains some general rules about how figures and tables are to be included in a written technical document. Note that these rules will automatically be relaxed somewhat for documentation that accompanies oral presentation materials since the person speaking can orally set the stage for the audience and does not need to rely on the documentation alone to convey meaning.

Thirteen general rules to use when you include graphics within documentation are as follows:

1. Always refer to the figure within the body of the text. Use one or two descriptive sentences. The citation sentence is generally of this form: "Figure 1 shows a schematic of the test setup used to perform the experiments" or "Table 2 includes data from the results of the post-tensioning tests."

2. Do *not* include a citation such as "As seen from the figure *below*" since with document editing and rearranging the figure may not appear "below" its citation. Similarly, do not include a citation such as "The *following* table includes data" because more than one table with data may follow the given paragraph.

3. Figures and tables should always be numbered within a document; however, figures and tables should be numbered separately (i.e., start numbering all figures from *1* and all tables from *1* within the same document). Numbering the tables and figures will enable you to cite them easily (see rule 1 in this list) within the body of the text since they will be easily distinguishable entities.

4. Generally, place the figure or table immediately *after* the paragraph that contains the citation sentence so that readers will then immediately see the figure or table referred to in the body of the text. This general rule has two notable exceptions:

    a. If the figure or table does not fit in the space immediately following its citation paragraph, include it at the top of the *following* page.

    b. In some cases, you may be given general instructions for the preparation of your document and be explicitly told to include all tables and figures as a separate document attached to the written document. This situation usually occurs when your document will be published professionally or in some cases when you are preparing a graduate thesis or dissertation.

5. Do not place a figure in a document *before* its citation sentence. Readers will stumble across the graphic out of context and not be able to determine its purpose within the document until later. Thus, your ideas will be poorly communicated.

6. Include enough white space on the page—both above and below the graphical image. If the text seems to flow directly into the image, readers will have difficulty discerning where the text ends and the image begins, and vice versa. To communicate effectively, you must avoid clutter on the page.

7. Always include a figure caption or table title with the graphical image. Figure captions consist of the figure number, often followed by a period (.), and its title; table titles consist of the table number, followed by a period, and a short title. Examples of acceptable figure captions and table titles include the following:

    a. Figure 1. Organizational Chart for Project Reporting

    b. Figure 34. Project Time Line

    c. Figure 12. Assembly Drawing of System Components

    d. Table 4. Experimental Data from Tension Test

    e. Table 6. Decision Matrix for Computer Purchase

8. Figure captions are generally included *below* the figure and centered within the text column; table titles are included *above* the table, also centered within the text column.

9. Figure captions and table titles are often given special treatment within the document, such as italics, bold, all caps, or differing font sizes. Avoid excessive use of these special features since it may detract from your ability to communicate effectively.

10. If you want readers to compare two figures or tables, you may refer to a previously cited graphical image by number within the body of the text. For example, your document could contain sentences such as the following:

Figure 5 shows the results from the tension test conducted with 15-mm-diameter specimens. When these results are compared with the results obtained from testing 20-mm-diameter specimens, shown in Figure 3, the following differences are noted.

In this case, Figure 5 would directly follow this citation, and Figure 3 would have been cited a few paragraphs before this citation.

11. In most cases, you will want to position the graphical image so that it is readable from the bottom of the page. Doing so enables readers to understand the image within the general flow of reading the document. If you have an exceptionally wide figure or table that will not easily fit vertically on the page, you may include it horizontally, as long as it is readable from the right-hand side of the upright page when the page is turned 90 degrees clockwise.

12. As always, if your graphical image contains part of a figure or tabular data that you obtained from another source, be sure to cite the source properly, as discussed in Chapter 3. Generally, the figure or table source is cited in the caption—for example,

    a. Figure 1. Organizational Chart for Project Reporting (Hamlin, 1993)

    b. Table 4. Experimental Data from Tension Test (Ahlborn, 2000)

Note, however, that for journal publications, you may be instructed to include the citation in a footnote instead.

13. Make sure your graphic is fully integrated into the document by adding one or more sentences *after* the figure that describe the data

presented in it. For example, your document might contain the following sentences in the paragraph immediately after the figure:

The data presented in Figure 3-4 clearly illustrate that the stress is proportional to the strain for materials in the elastic range. The straight-line fit through the data has a coefficient of determination of 0.9, which indicates the data are well correlated. Results portrayed in this figure are also an indirect indication of the validity of the experimental method used in the testing procedure.

9.6 Effective Use of Image Files

Function. Although simple figures and charts can typically be created from within your word processor, you will often need to work with externally created figures by importing them into the document as needed. Since your ultimate goal is to have clear, distinct figures in your document that illustrate your work and enhance your document, you must learn the differences between file types and how they are used effectively in technical communication.

Form. The two basic types of image files are vector files and bitmapped files. Each type of file is useful in its own right; however, make sure you know the difference between them so that you obtain the result you desire.

Vector files contain graphics stored as mathematical descriptions (i.e., vectors). Since vector images are stored independent of the resolution of the input or output device, they are scalable within the document. Thus, if you enlarge or shrink the size of the image within the document, its quality will remain unchanged. Vector graphics are generally preferred for printed documents.

Bitmapped images are stored as pixels or dots. A bitmapped image will look good only if it is printed at about the same resolution and same size at which it was created. Bitmapped images do not scale well; the pixels often become jagged in the final documents. A bitmapped image is preferable for use in online documentation such as a Web page, but it might not work well within a printed document. Bitmapped images are highly dependent on resolution, or dots per inch (dpi). In general, a low-resolution bitmapped image is best for online documentation and a high-resolution bitmapped image is necessary for print.

As with document files, image files have been standardized. The four most commonly used graphical file types are discussed next.

Encapsulated PostScript. PostScript is a vector-based graphical language. An *encapsulated PostScript (EPS, or .eps)* file contains the vector-based data as well as bitmapped data for display on the computer screen. Thus, if you use EPS files in your documents that are to be printed, you have the best of both worlds: high-quality print from the vector-based data and easy-to-display raster data on your computer screen as you are working on your document. For print documents, EPS files are probably the best type of image file you can use for clarity and versatility. You may encounter problems when working with EPS files if you have created them for use on one platform and migrate them to a different one. Remember that the raster image embedded on top of the PostScript file is for display purposes only, and files that display well on a Macintosh might not display well on a PC. However, since the vector-based PostScript file is in the background, the image will usually print out at high quality without a problem, even if the image on the screen seems distorted. Some file conversion programs will take a bitmapped file such as a TIFF file (see the following section) and convert it to an EPS file. If this happens, the bitmapped file becomes vectorized in the background and may not print out well at different resolutions.

Tagged image file format. *Tagged image file format (TIFF, or .tif)* files are the most widely used type of graphical file for all kinds of technical communication. A TIFF file is a bitmapped raster image that can encode gray scale, RGB (red-green-blue), or indexed color. Because it is one of the most commonly used file types, the TIFF file is typically readily transportable between software applications. Because TIFF files are raster images, you will need to think about image resolution when creating the file. A low-resolution file will be acceptable for on-screen documents but may print poorly. In general, TIFF files will print well only if they are printed at the same resolution at which they were created. You also need to consider the relative size of a TIFF file: the image should be created at approximately the same size that it will be printed since TIFF files may not scale well.

Graphic interchange format. The *graphic interchange format (GIF, or .gif)* file format is used extensively online in documents such as Web pages. This file type is usually relatively small, so on-screen display times for the graphical images are manageable. GIF files can be readily edited and modified so that images created elsewhere can be altered to suit your needs. One disadvantage of GIF files is that they use *indexed color,* which means the palette is limited to 256 colors. Photographs and scanned images can be saved in GIF format as long as they are reduced to 8-bit indexed color. GIF files use a *lossless* compression scheme, which means that if the file size is compressed, no information is lost. Thus, files can be compressed and uncompressed with no loss in image quality.

Joint photographic experts group. The *joint photographic experts group (JPEG, or .jpg)* image file type was designed to be used with continuous-tone photographs. This file type uses millions of colors and is not limited to 256 indexed colors as GIF files are. JPEG files are used widely on the Web, but you should ensure your file size does not become too large, which will make them take too long to download or display. You would typically not use JPEG files for line art or for images with large areas of flat color since they will not compress well in this format. JPEG files use a *lossy* compression scheme, which means you will lose information as a file is compressed. If you compress and uncompress a JPEG file several times, the result will probably be an image that bears little resemblance to your original.

Mechanics. You must think about the final destination of your image file before it is created. Since image files typically cannot be easily manipulated, you should save your work in its native format for later editing. For example, if you create a drawing in AutoCAD that you want to bring into a document, make sure you save the file as a TIF or an EPS file, as well as in its native .dwg format. This way, if you find errors in your drawing, you can go back into AutoCAD to make the corrections and then write out a new TIF or EPS file for your document.

For documents you intend to print, choose an EPS file for line art or a TIF file for images with continuous colors or gray scale. If you are creating TIF files, make sure your resolution is set at approximately the same as your printer's resolution (most standard printers are set at 300 or 600 dpi). For online documents, choose a much lower resolution, which will not affect their on-screen appearance and will save a lot of time in their display. Another consideration is that online colors may look different across platforms: color images created for viewing on a Macintosh will often appear to be different shades of colors when displayed on a PC, and vice versa. Usually this difference is not a problem; however, in some cases, you may want to choose different color schemes to achieve your desired results across platforms.

9.7   Exercises

1. For an airplane to maintain a steady, level flight for a given airspeed, the available power provided by its engines must exceed the required power. The power available to the airplane is also a function of its airspeed since the propeller efficiency varies with airspeed. The following table supplies data for a particular aircraft:

Airspeed (mph)	60	70	80	90	100	110	120	130	140	150
Power required (hp)	53	51	52	55	61	70	82	96	113	134
Power available (hp)	90	99	105	110	114	117	120	121	122	121

Graph these data and determine the maximum airspeed for the aircraft. Make sure you fit the curve through the data points and do not just connect the dots. Be sure to include all appropriate labeling. Write a memo to your instructor, including your findings and the graph.

2. As a fluid flows through a pipe, it experiences an energy decrease, caused by interaction between the fluid and the pipe walls, that is known as *head loss*. The following table includes data for head loss in iron pipes as a function of water velocity:

| Velocity (ft/s) | Head Loss (ft/1,000 ft) | |
	4-In. Pipe	6-In. Pipe
0	0	0
1	1.6	0.7
2	4.2	2.4
3	8.2	4.7
4	13.5	7.7
5	20	11.4
6	28	15.5
7	37	20

Graph these data and turn the graph in to your instructor. Make sure you fit a smooth curve through the data points and include all appropriate labeling.

3. Fuel consumption in a car is related to speed and road grade (slope). The following table contains data from a test conducted at a standard 5 percent road grade for a particular car model:

Speed (mph)	Fuel Rate (gal/h)
32.2	1
54.4	2
68.4	3
78.1	4
84.8	5

Create a bar chart illustrating these data, including appropriate labeling, and turn the chart in to your instructor.

4. Estimate the number of hours you spend each day engaged in the following activities: sleeping, attending class, studying, exercising, eating, and relaxing. Create a pie chart showing these data. Meet within a small group to compare and discuss your results. Are significant differences in time allocation apparent among your group members? Could these differences lead to conflicts among team members? Turn in a memo to your instructor summarizing your discussion and attaching all pie charts.

5. Locate three graphics on the Web that you think are misleading. Write a memo to your instructor indicating what you think the main problems are with each. Turn in a hard copy of the graphics along with your memo.

6. Create an organizational chart for a student group to which you belong. Create what you think is an improved structure for the organization. In a memo to your instructor, describe why you think your structure is better than the existing one. Attach both diagrams to your memo.

7. Revise the visuals you created in Exercise 2 of Chapter 8. Include at least two graphical images within your set of presentation visuals.

8. Make a scatter plot of the data shown next, using spreadsheet software. These data represent the measured annual discharge (X) and estimated annual recharge (Y) of an Artesian aquifer. The units for the data are acre-feet in thousands. Determine the fitted line for the data when Y is the dependent variable. Make sure you include labels, units, and so forth on your plot. Show the line, the equation of the line, and the coefficient of determination on the plot. In a memo to your instructor, discuss whether you believe X is a good predictor of y (i.e., whether the fitted line is a good representation of the data). Attach the scatter plot to the memo.

X	Y	X	Y
11.3	11.5	15.4	14.1
12.8	13.6	9.8	9.6
12.9	14.8	15.2	14.4
16.5	14.7	11.9	11.7
10.5	10.1	14.1	14.6
16.5	16.3	13.0	12.8
10.8	9.8	13.6	12.5
16.0	17.3	11.1	10.4
11.6	12.5		

9.8 Additional Resources

Bertoline, G. R., E. N. Wiebe, and C. L. Miller. 1998. *Fundamentals of graphics communication.* Boston: WCB/McGraw-Hill.

Cyganski, D., and J. A. Orr, with R. F. Vaz. 2001. *Information technology inside and outside.* Upper Saddle River, NJ: Prentice Hall.

Earle, J. H. 2004. *Engineering design graphics.* Upper Saddle River, NJ: Pearson Education.

Giesecke, F. E., A. Mitchell, H. C. Spencer, I. L. Hill, R. O. Loving, J. T. Dygdon, and J. E. Novak. 2004. *Engineering graphics.* Upper Saddle River, NJ: Pearson Education.

10

Communication for Job Searches

At some point in your career, likely more than one point, you will need to find a job. Whether the time is directly before or after graduation or later in your career, this chapter should help you with that search. Experts have differing opinions about some areas of job search communication; we discuss these opinions so that you can better decide how you want to approach such issues.

10.1 Résumés

Function. A *résumé* gives a potential employer your background so that she can decide whether to consider you further for a position. Since you probably don't want to be a permanent fixture on a university campus, being able to construct an appealing résumé is essential as you begin your career. Likewise, chances are, you will not stay with your first job throughout your working career, so you will also need to be able to write résumés periodically as you advance within a company or as you migrate to different companies.

Form. A résumé should be concise yet include all information necessary to describe your background. It must also be honest and completely error free. Typical sections of a résumé are as follows:

Identifying information. At the top of the résumé, centered on the page, you should put your name, address, telephone number or numbers, and e-mail address. Do not include your e-mail address if you are concerned that your current employer might see an e-mail from your prospective employer. Students just graduating may want to put their name at the top, centered, and then place their current address and permanent address below. The current address should be placed at the left margin and the permanent address at the right margin. Doing so is especially important if you are about to graduate and the company might try to contact you after you leave campus.

Objective. In the Objective section, you should provide a one-sentence statement that describes the kind of job you seek. The statement should not be too specific, particularly for a recent graduate; however, it should not be too general, either. If you want to work in research and development, state this: don't merely say you want a challenging job as a mechanical engineer. You don't want to limit your chances too much, but at the same time you don't want to appear as if you have no idea about what you want to do. Most people use an objective statement on their résumé, but it is optional. If you want to use your résumé to apply for several jobs for which your objective would change, do not put your objective in your résumé. Instead, include it in your cover letter (described subsequently).

Education. In the Education section of your résumé, you describe your postsecondary education. Do *not* put secondary education (high school) information in this section. Your education should be given in reverse chronological order (i.e., most recent degrees first). You should include the degree, the date it was (or will be) awarded, and the name of the university or college. The address of the educational institution is optional and typically not included, but if you do include it, use only the city and the state. A recent graduate may want to expand the Education section to include such items as minors, relevant coursework, grade point average, and specific computer skills if he is particularly adept at them.

Experience. In the Experience section, you describe your work or professional experience. It should also be in reverse chronological order, with your most recent job listed first. For each job you list, include the dates of employment, the employer's name, your job title or position, the employer's address (city and state are sufficient), and a brief description of your duties on the job. If the position you are applying for is your first "engineering" job, don't negate the positive aspects of a job you may have had working in a fast-food restaurant or some other nonengineering company. If you were shift manager, this fact implies that you could handle responsibility and that your supervisor viewed your work positively. Engineering managers seeking permanent or summer employees look for individuals who have distinguished themselves and shown an excellent work ethic—even if it was only for flipping burgers. Note that for people with significant work experience who have been out of school for a number of years, placing the Experience section before the Education section is common and recommended. Some people believe the Experience section should always be listed first. We suggest you put your strongest assets first, and for a recent graduate, your education is likely your strongest asset.

Other information. Information such as activities in college, professional affiliations, and hobbies may be included if you have a good reason to do so. You should never include personal information such as marital status, children's names, or a photograph; however, in some cultures, this information is expected, so if you are applying for a job in a foreign country, you should try to determine whether such information is necessary. Also never include a statement such as "Health: Excellent" since *excellent* and *very good* are the only descriptors people ever use: you can put the space to better use. You may include height, weight, and date of birth, but we do not recommend you do so. Companies don't want to know this information, which could be used against them in allegations of discrimination. If you have published papers or books, you should put this information in your résumé. Any academic or work-related honors or citations should be included, and scholarships should also be mentioned. Military service, if any, should appear in a separate section. If you speak a foreign language, be sure to state this fact on the résumé.

References. Usually you should limit the length of your résumé to one or two pages, and including full contact information for three references can take up a lot of space. For this reason, the résumé should simply include a statement such as "Upon request" or "Available on request" in the References section. You should have a list of three to five people who have agreed to serve as references available, but do not include it with your cover letter and résumé unless the organization to which you are applying has specifically requested you to do so. To include references or not is one area of differing opinions. Some people recommend including three to five references on a separate sheet. The reason is twofold: First, your prospective employer can seriously consider you and conduct a reference check without contacting you. This possibility is beneficial to your prospective employer because it saves time, particularly if you have already been in contact with her prior to sending her your résumé. Second, including references can make your prospective employer believe you have confidence in what your references will say about you, and a references list shows that you have taken the time to contact references. You must decide which approach you prefer.

Mechanics. As mentioned previously, a résumé should be one, or at most, two pages long unless you have significant work experience that requires a fairly in-depth description or you have a significant number of publications. Résumés for recent graduates typically should not exceed one page. References, if included, may be placed on a separate sheet. If your résumé exceeds one page, print each page on a separate sheet and staple the pages together in the upper right corner. Do not print your résumé on both sides of a sheet of paper since doing so makes your résumé more difficult to read. Use a standard-size, 8½ × 11-inch, high-quality bond paper so that your résumé does not get lost (if smaller) or become difficult to handle (if larger). Use one or at most two font types and print the document on good-quality white or light colored paper. Don't use colored paper because doing so may make your résumé more difficult to read. Last, make good use of white space to help make your résumé readable. *Remember:* Your résumé is your chance to make a first impression on a potential employer. A sloppy résumé will likely mean you are not invited for a job interview, since the message you will be conveying is that you really didn't care enough to spend the time putting together a neat résumé. The implication is that you also won't care enough about the job you will be doing for him.

The mechanics for each section of your résumé follow. In this discussion, we refer to two résumés: Figure 10.1, which represents the résumé of a graduating engineering student, and Figure 10.2, which represents the résumé of an experienced engineer.

Identifying information. Figures 10.1 and 10.2 show two methods for including identifying information on your résumé. Figure 10.1 shows the case with two addresses, one of which is a permanent address, and Figure 10.2 shows only one address—most likely the person's permanent address. Note that no heading is included with this information.

Objective. Your job objective can be worded in terms of the specific job position or its functions. Figure 10.1 shows the job function option, and Figure 10.2 shows the job title option. If you word your objective too narrowly, you will limit your options. For instance, the objective "An entry-level mechanical engineering position in the automobile industry" would limit you to the automobile industry. As a recent graduate, such an objective would be appropriate only if you wanted to place such limits on your job possibilities. *Remember:* This section is optional, but most people include it. This section could be titled "Objective," "Job Objective," or "Career Objective."

Education and experience. As stated previously, your education and work experience should be described in reverse chronological order. For a student who has recently graduated or is about to graduate, we suggest you list your education first (Figure 10.1). Since you will likely have only one degree, you may want to include some other information, such as minors, relevant coursework, grade point average, and specific computer skills, in this section. Figure 10.2 shows an example for an engineer with many years of experience. Note that the Experience section is listed *before* the Education section. In Figure 10.2, in the Education section, the two degrees are listed in reverse chronological order, and the educational information is not expanded.

Four rules pertain to describing work experience:

1. Be brief. Use short phrases rather than complete sentences.
2. Do not use the first person, *I,* or the second person, *he* or *she.*

William M. Bulleit

College Address
833 College Avenue
West Lafayette, IN 47907
Telephone: 765-555-5555
Email: wbulleit@purd.edu

Permanent Address
521 Spruce Lane
Leola, PA 17540
Telephone: 717-555-5555
Email: wmbulleit@eplace.net

OBJECTIVE

    Civil engineering position that emphasizes structural engineering

EDUCATION

    Purdue University, West Lafayette, IN
    B. S. Civil Engineering, May 2004, GPA: 3.49
    Structures emphasis. Courses: Structural Analysis, Matrix Structural Analysis,
    Reinforced Concrete Design, Steel Design, Timber/Masonry Design.

EXPERIENCE

    Chicago Bridge & Iron, Inc., Oakbrook, IL, Summer 2003, Engineer
    Performed analysis and compilation of structural reports for hydrocracker reactors.
    Assisted in analysis of water tanks.

    H. M. Stauffer and Sons, Inc., Leola, PA, Summer 2001 & 2002, Worker
    Fabricated prefab wood stud walls and metal-plate-connected wood trusses.

HONORS

    Chi Epsilon Honor Fraternity
    Tau Beta Pi Honor Society

ACTIVITIES

    American Society of Civil Engineers: Student member
    Concrete canoe team: Member. Worked on design and fabrication.

REFERENCES

    Will be provided on request.

Figure 10.1. Sample résumé for a graduating engineering student

3. Use the appropriate tense to describe your work. Your present job should be described in present tense and your past jobs in past tense, unless you have been at your present job long enough that you had past responsibilities. Figure 10.2 shows an example of the use of past tense in the description of the present job.
4. Use active verbs such as *accomplish, implemented, maintain,* and *designed.*

William M. Bulleit
521 Spruce Lane
Leola, PA 17540
Telephone: 717-555-5555
Email: wmbulleit@eplace.net

Career Objective: To become chief structural engineer for a medium to large architectural/engineering firm

Work Experience:

1990 – Present

Structural Engineer
Wood Engineers, Inc., Lancaster, PA
Responsible for structural analysis and design of a wide range of wood structures, including glulam arches, wood trusses, traditional timber frames, and wood shear walls and diaphragms. Design one- and two-story steel frames and concrete footings. Performed research on reinforced wood and traditional timber frames.

1980 – 1990

Structural Engineer
HNTB, Bellevue, WA
Worked on the following projects:
West Seattle Freeway Bridge design check. Post-tensioned box-girder bridge built by the balanced cantilever construction method. Performed superstructure design.
Mt. Baker Ridge tunnel on I-90 in Seattle. Soft earth tunnel using multiple drift construction, inside diameter approximately 62 feet. Analyzed tunnel opening behavior and designed access pits.
I-210/I-10 Interchange ramp, Lake Charles, Louisiana. Horizontally curved plate-girder bridge with R/C slab. Performed substructure and superstructure design.

1976 – 1980

Structural Engineer
Perry Oceanographics, Inc., Riveria Beach, FL
Performed analysis and design of submarine hulls and diving bells, both with diver lockout capabilities. Supervised hydrostatic testing of fabricated hulls. Compiled structural reports for registration of submersibles.

1974 Summer

Engineer
Chicago Bridge & Iron, Inc., Oakbrook, IL
Performed analysis and compilation of structural reports for hydrocracker reactors.

Education:

Purdue University, West Lafayette, IN
MSCE, December 1975
BSCE, May 1974

Professional Activities:

Registered Professional Engineer in Washington and Pennsylvania
Member, American Society of Civil Engineers (ASCE)
Member, Timber Framers Guild
Member, Prestressed Concrete Institute

Journal Publications:

Bulleit, W. M., Sandberg, L. B., Drewek, M. W., & O'Bryant, T. L. (1999). Behavior and modeling of wood-pegged timber frames. *Journal of Structural Engineering,* ASCE, *125*(1), 3–9.

Figure 10.2. Sample résumé for an experienced engineer

Bulleit, W. M., Sandberg, L. B., & Woods, G. J. (1989). Steel reinforced glued laminated timber. *Journal of Structural Engineering,* ASCE, *115*(2), 433–444. No. 2, 1989, pp. 433-444.

Conference Proceedings:

Bulleit, W. M., & Drewek, M. W. (1999). Analysis of wood-peg connected timber frames. In R. R. Avent & M. Alawady (Eds.), *Structural engineering in the 21st century: Proceedings of the 1999 Structures Congress, New Orleans, LA, April 18–21, 1999* (pp. 654–657). Reston, VA: American Society of Civil Engineers.

Bulleit, W. M., Sandberg, L. B., O'Bryant, T. L., Weaver, D. A., & Pattison, W. E. (1996). Analysis of frames with traditional timber connections. In V. K. A. Gopu (Ed.), *Proceedings of the International Wood Engineering Conference* (Vol. 4, pp. 232–239). Madison, WI: Omnipress.

Bulleit, W. M. (1993). Wood combined with other materials. In D. A. Bender (Ed.), *Wood products for engineered structures: Issues affecting growth and acceptance of engineered wood products* (pp. 44–46). Madison, WI: Forest Products Society.

Workshops Attended:

Wind Engineering, Multiprotection Design Summer Institute, July 29 - Aug. 2, 1985, Emmitsburg, MD, Sponsored by FEMA, NSF, USGS.

Earthquake Protection Design, Multiprotection Design Summer Institute, July 16-19, 1984, Emmitsburg, MD, Sponsored by FEMA, NSF & USGS.

References: See attached sheet.

Figure 10.2. (*continued*)

In Figures 10.1 and 10.2, two approaches to supplying this information are shown. In one approach, the company name is shown first, followed by the job title; the second approach reverses the order of these two items. The choice of whether to show the company name or the job title first is your decision. Typically, whichever is more impressive should be given first, but once you make your choice, be consistent throughout the résumé.

Other information. Figures 10.1 and 10.2 show some additional information and ways it can be included. The best guideline for whether you should include a specific piece of information on your résumé is this: present the information if you think it will help you get the job. For students about to graduate, your résumé should include information about your university experiences simply because you don't have much to put on it. Later in life, after you establish a career track record, this type of information is not normally included on your résumé. For example, the résumé shown in Figure 10.1 illustrates the inclusion of information about extracurricular activities, and the résumé shown in Figure 10.2 does not, although it does show workshops related to professional development.

References. Use a statement similar to that shown in either Figure 10.1 or Figure 10.2 in the References section of your résumé. Be sure to have available a sheet listing three to five people who have agreed to provide you with references.

You should have previously contacted these individuals to obtain their permission and to let them know they may be contacted in the future. A good idea is to give each of your references a copy of your résumé—especially if one of your references is a university professor who may not know enough details about you without a copy of your résumé. Remember that including your list of references with your résumé and cover letter is acceptable, and sometimes recommended. If you do include your references, remove this section from the résumé and put a statement in your cover letter about the included references, or leave this section in with a statement such as "References attached" or "References included on a separate sheet."

Examples

Lewis (1988), although a bit dated with respect to electronic communication, is a good reference for examples of résumés. The book includes about 50 sample résumés that cover a broad range of possibilities. The examples in Figures 10.1 and 10.2 have a *chronological organization* (i.e., your experience is organized chronologically). Another technique is referred to as *analytical organization*. In this case, your experience is organized by skills (e.g., technical skills, laboratory skills, and management skills). This type of organization is primarily useful for a person who has a broad range of skills and wants to highlight them.

Résumé Checklist. Following is a checklist for résumé development:

- ✓ Did you keep your résumé brief: one or two pages maximum?
- ✓ Did you use 8½ x 11-inch high-quality bond paper of a pale color?
- ✓ Did you place your name, address, telephone number(s), and e-mail address at the top?
- ✓ If you listed a job objective, is it appropriate for your education and experience?
- ✓ Did you list your education in reverse chronological order?
- ✓ Did you include the dates of all degrees received?
- ✓ If you are a recent graduate, did you include some extra educational information?
- ✓ Did you *omit* high school information?
- ✓ Did you list your work experience in reverse chronological order?
- ✓ Did you use action verbs and implied pronouns when describing your work experience?
- ✓ Did you include academic and work-related honors and citations?
- ✓ If applicable, did you include military service?
- ✓ If you chose to, did you include activities and hobbies?
- ✓ If you speak a foreign language, did you say so?

✓ Did you include a statement that references are available on request?

✓ Did you contact and develop a list of references for future use as necessary?

✓ Did you *omit* personal information such as your spouse's name, your children's names, and a photograph of yourself?

✓ Did you make your résumé visually appealing by using wide margins and enough white space?

✓ Did you minimize the number of font type?

✓ If your résumé is longer than one page, did you staple the sheets together?

✓ Did you proofread your résumé, looking for misspellings, typos, and grammatical errors?

✓ Did you proofread it a second time?

10.2 Cover Letters

Function. The *cover letter* is included with your résumé and introduces you to the person reading your résumé. The cover letter provides you with the opportunity to expand on information in your résumé and to include some information you believe is important but does not fit on your résumé. The cover letter also gives you the opportunity to demonstrate to your potential employer that you can write clearly and in a grammatically correct manner. If your résumé does not include an Objective section, you should state your job objectives in the cover letter. The ultimate goal of the cover letter and the résumé is an invitation to a job interview. Dealing with a job interview is discussed in Section 10.4.

Form. A cover letter has the form of a letter but must adhere to a few specific "rules." It should have no more than four or five paragraphs and should not exceed one page. The first paragraph of the cover letter should state why you are writing to the particular company or individual, and you should state how you heard about the job opening. The middle one to three paragraphs should highlight important parts of your résumé and expand on them as you see fit. If you did not include a job objective in your résumé, your objective should be stated in the second paragraph of your cover letter. The last paragraph of the cover letter is a concluding statement in which you thank the reader for her consideration and say you hope she is interested in hiring someone with your abilities. If appropriate, state that *you* will call her later as a follow-up to the letter and résumé.

Mechanics. The mechanics for a cover letter are the same as for a standard letter, although a few items apply specifically to a cover letter. You should, if at all possible, address the cover letter to a specific person in the company. Taking the time to determine who should receive your résumé can give you an edge in your job search. If you are responding to an ad or another form of job posting and the supervisor's name is not available, address the letter to the title of the person, which is likely shown in the ad. Use "To Whom It May Concern" only as a last resort. Don't use "Dear Sir" or "Dear Madam" unless you are sure you will not offend anyone with either of those terms. In the final paragraph of the cover letter, a statement like "Thank you for your consideration and I hope to hear from you in the near future" is reasonable. A statement like "I will follow up this letter with a call in a week or so" is also reasonable if you think an active approach is appropriate. An active approach can be beneficial—if you do not appear to be too pushy. Figure 10.3 shows an example of a cover letter for responding to a job posting that lists a contact person. This cover letter was developed to go with the résumé shown in Figure 10.1. Note that the person who wrote the letter took the time to learn some specifics about the company. Knowledge about specific work the firm does shows an interest in the firm and allows the applicant to demonstrate how he might fit into the company.

Examples

Lewis (1988) includes examples of 12 cover letters covering a range of possible types.

---

833 College Avenue
West Lafayette, IN 47907

10 February 2004

Mr. David Anderson
Structural Design Consultants, Inc.
Southfield, MI 48034

Dear Mr. Anderson:

I recently learned from a posting here at Purdue University that your firm has an opening for an entry-level structural engineer. I have attached a copy of my résumé so that I can be considered for this position.

As shown on my résumé, I have emphasized structures in my undergraduate work and expect that I could perform well at your firm. I have worked with steel structures at Chicago Bridge & Iron. Although that work involved primarily pressure vessels, I also used the AISC steel design code to design supporting structures. My experience fabricating wood walls and trusses helped give me insight into wood structures that was reinforced in my timber design studies at Purdue. I know your company does some wood design, particularly for large residential structures, and I'm confident that I could perform well with that type of design.

I look forward to hearing from you and hope we can set up an interview in the near future. If you contact me after May 16, 2004, please use my permanent address shown on my résumé.

Sincerely,

William M. Bulleit

Figure 10.3. Sample cover letter for a graduating engineering student

10.3 Documents for Electronic Job Searches

Several issues related to the use of e-mail and the Web to search for jobs need to be discussed. The goal is to get a job, and the résumé and, sometimes, the cover letter are still needed. The difference between an electronic job search and a traditional job search is that the information will be sent to your prospective employer by means of electronic media or will be sent in printed form and then scanned into electronic format. Your prospective employer may also use electronic techniques to screen résumés received from applicants. If this is the situation, you will need to incorporate keywords into your résumé to get it beyond the electronic screening.

Function. The function of an electronic job search is similar to that of job searches through traditional means: to get a job interview and ultimately to gain employment.

Form. The three primary forms of electronic job searches are as follows:

1. Your prospective employer requests that you send your résumé by e-mail. In this case, your "cover letter" is the body of the e-mail, and you attach your résumé to the e-mail or put it in the body of the e-mail. The choice you make depends on what the company you are applying to asks you to do and how your résumé will be treated (e.g., read electronically).
2. Your prospective employer requests a printed résumé to be scanned into a database.
3. You place your résumé on the Web.

Mechanics. Consider each of the preceding three situations:

1. A number of possibilities exist for sending a résumé by e-mail. The easiest is when the prospective employer intends to print out the résumé for someone to read. The employer will likely request that you send the résumé in a specific format: in specific word processor software format; *rich text format (RTF, or .rtf),* which is readable by many word processors; *portable document format (PDF, or .pdf);* or ascii text format. Of these, only the last will require you to modify the form of the résumé. If one of the first three is requested, the file should be attached to your e-mail message. Your e-mail message should then be the "cover letter," unless your prospective employer requests you to keep the e-mail message short. Then do so, but make sure your résumé is complete since you will not be able to include additional information in the cover letter. Do not include

references in an electronic résumé unless you are specifically requested to do so.

If your prospective employer requests that the résumé be in ascii text format, she likely has two reasons: she wants you to place the résumé in the body of the e-mail or she wants to search a large number of résumés electronically. Whenever a text format is requested, assume that the résumé will be searched electronically, unless you know otherwise. Use wide margins and left justify the text. The wide margins help guarantee that the electronic reader reads all the information on the résumé. Use a single-column format (we do not discuss the double-column format), and when typing your résumé, use the space bar to move horizontally since the tabs on the electronic reader may not be the same as yours. To check your text résumé, save it as a text file and open it in a simple text editor. Any characters that appear incorrectly need to be replaced. Furthermore, if your file will be read electronically, be sure to include keywords, or industry-specific jargon, related to the position. For instance, if the job you are applying for requires knowledge of certain software packages, you need to put the names of such software in your résumé. If the position requires you to speak in public, use the words *public speaking* or *oral communication* in the résumé. You will need to give some special thought to your résumé if it will be searched electronically. The résumé shown in Figure 10.1 is shown in text format in Figure 10.4.

2. If the company will scan your printed résumé into a database, you need to think about how well your résumé will scan. You should use sans serif fonts such as Arial or Helvetica since these scan better. Avoid special characters, underlining, italics, and so forth. Use wide margins to help guarantee that the scanner reads all the information on the résumé. Use a single-column format. Also include keywords, just as you would for a text file that will be read electronically, since the scanned file will be searched electronically.

3. The specific mechanics of putting your résumé on the Web are beyond the scope of this text, but we mention a few important points. The résumé will need to be developed in *hypertext markup language* (*HTML, or .html*). You will also want to include keywords in it so that companies searching the Web will be more likely to find your résumé. You can place the universal resource locator (URL) for your Web site in an e-mail to a prospective employer, indicating that he can look for your résumé there. If you do this or if you choose to reference your personal Web site in your résumé, be careful how much unnecessary information appears on your site. If you typically place personal photos on your site or discuss politically or socially controversial issues on your site,

William M. Bulleit

College Address	Permanent Address
833 College Avenue	521 Spruce Lane
West Lafayette, IN 47907	Leola, PA 17540
Telephone: 765-555-5555	Telephone: 717-555-5555
Email: wbulleit@purd.edu	Email: wmbulleit@eplace.net

OBJECTIVE

Civil engineering position that emphasizes structural engineering

EDUCATION

Purdue University, West Lafayette, IN
B. S. Civil Engineering, May 2004, GPA: 3.49
Structures emphasis. Courses: Structural Analysis, Matrix Structural
Analysis, Reinforced Concrete Design, Steel Design, Timber/Masonry Design.

EXPERIENCE

Chicago Bridge & Iron, Inc., Oakbrook, IL, Summer 2003, Engineer
Performed analysis and compilation of structural reports for hydrocracker
reactors. Assisted in analysis of water tanks.

H. M. Stauffer and Sons, Inc., Leola, PA, Summer 2001 & 2002, Worker
Fabricated prefab wood stud walls and metal-plate-connected wood trusses.

HONORS

Chi Epsilon Honor Fraternity
Tau Beta Pi Honor Society

ACTIVITIES

American Society of Civil Engineers: Student member
Concrete canoe team: Member. Worked on design and fabrication.

REFERENCES

Will be provided on request.

Figure 10.4. Résumé in text format

directing a prospective employer there is not a good idea. Thus, if you plan to put your résumé on your Web site, keep your site clear of links and information that might reflect poorly on you. For more information, search the Web with keywords such as *electronic resumes, employment, jobs,* and *resumes.*

10.4 Interview Communication

Function. During the interview, you must communicate with the interviewer in a manner that makes the best impression. The interviewer will expect you to be nervous, so perfection is not required; however, you should be as composed and articulate as possible.

Form. Interview communication is primarily oral, but your body language, the way you dress, and your attitude will also have an impact on the overall impression you make. Try to speak in complete sentences and to anticipate questions that may be posed. Your university career center will likely be a good source of information for interview tips. When asked a question, pause to reflect on your answer: if you answer too quickly, you may make mistakes.

Mechanics. In preparation for an interview, learn as much as you can about the company. Then, go over your résumé and be ready to answer questions about it. The interviewer may want you to expand on certain areas in your résumé. Be prepared to go into details and do not, under any conditions, exaggerate. You will almost certainly be asked why you want to work for the company where you are interviewing and why you want the specific job you are applying for. Be ready for these questions. Also be prepared to discuss your long-term career goals. In answering all questions, be brief and direct. You do not need to give your life history.

Nonverbal communication during your interview is also important. Dress appropriately: a suit for men, and a suit, skirt and blouse, or dress for women. Dress conservatively (e.g., dark colors, no flashy jewelry, simple tailoring, and no bow ties). Show an active interest in the interviewer and the company. Sit up, don't slouch, and maintain good eye contact with the interviewer. Do not try to be someone you are not. You do not know what type of person the company is looking for, so just be yourself. The interview is just as important for you as it is for the company representative. Ultimately, you do not want to work for a company where you will be unhappy, so take the time during the interview to find out what the working environment is like to determine whether the company is a place where you would like to work. Do not respond in anger if the interviewer pushes you. She may be testing you to see how you respond under pressure. This approach is not likely, but possible. Try not to be nervous about being nervous. Interviewers expect some nervousness. Finally, be a good listener and ask appropriate questions. Do not ask about such things as vacations, holidays, sick days, and retirement benefits. Discuss these benefits only when the interviewer broaches the topic. Last, do not ask too many questions about where the job might lead in the future. Remember that the company needs someone for the immediate job. If you think a question is appropriate about the future of the job, phrase the question something like this: "If I were to get this job and perform well, what sort of responsibilities might I expect in 5 years?" This question shows that you clearly see the need to perform well in the job you are applying for, but you are interested in your future with the company.

Interview Checklist. Following is a checklist for interview preparation:

✓ Did you take time to learn about the company?
✓ Do you know some specifics about the job you are applying for?
✓ Are you prepared to expand on your résumé if necessary?
✓ Have you thought about your long-term career goals?
✓ Have you recently reread your résumé?
✓ Have you thought of specific questions you want to ask?
✓ Are you dressed appropriately?
✓ Even though you are nervous, are you confident you can make a good impression and keep a positive attitude?
✓ Are you truly interested in the job and the company?
✓ Are you prepared to listen carefully and to ask appropriate questions?

10.5 Postinterview Communication

Function. The purpose of follow-up communication is to thank the company for interviewing you and to reiterate your interest in the company. A hidden purpose is to remind your interviewer of your abilities and strengths.

Form. A follow-up letter after an interview takes the form of a letter. It should be short and addressed to the person who interviewed you.

Mechanics. The mechanics are the same as for a standard letter, although a few items apply specifically to a follow-up letter. The letter should consist of no more than two paragraphs. In the first paragraph, thank the interviewer for the interview opportunity. Include the date of the interview to help the interviewer remember you in case he had several interviews during a number of days. You should close the paragraph by reiterating your interest in the company. The second paragraph is a place to say that you will call the company to follow up on your interview, but if you say this, you *must* call. If you do not feel comfortable with the active approach or the interviewer told you specifically not to call, omit the second paragraph and, by all means, do not call.

10.6   Exercises

1. Write a résumé for yourself. Have a faculty member or an appropriate professional critique it for you.

2. In a group of three or four students, examine and discuss each of your résumés. Each of you should then rewrite your résumé and the group should gather again to discuss the rewritten résumés.

3. Do Exercise 1 for a cover letter. Write the cover letter to a company to which you expect to apply or would like to apply.

4. Do Exercise 2 for a cover letter.

5. Hold a mock interview. The best approach is to have a professional interviewer conduct the mock interview, but anyone with interview experience will suffice.

6. Do Exercise 1 for a postinterview letter.

7. Do Exercise 2 for a postinterview letter.

8. Make your own checklist for writing a cover letter.

10.7   References

Lewis, A. 1988. *The best resumes for scientists and engineers.* New York: Wiley.

Appendix: Grammar

Basic Punctuation

Apostrophes to form contractions. Use an apostrophe to form contractions such as *I'm, you're, can't,* and *don't.*

Apostrophes to form plurals. Use an apostrophe to form the plural of single letters and numbers.

Wrong: You have too many As in the title.

Correct: You have too many A's in the title.

Do not use an apostrophe when more than one letter or number is involved.

Wrong: Prestressing steel was introduced in the 1920's.

Correct: Prestressing steel was introduced in the 1920s.

Apostrophes to form possessives. An apostrophe is used to make the possessive form of a noun.

Wrong: The boss' office has been moved.

Correct: Tom's calculator is on the desk.
Correct: The boss's office has been moved.
Correct: The engineer's decisions were wrong. (One engineer)
Correct: The engineers' decisions were wrong. (More than one engineer)

The possessive form of some personal pronouns does not require an apostrophe (i.e., *hers, its, ours, theirs,* and *yours*). The possessive form of indefinite pronouns requires an apostrophe.

Wrong: The computer is her's.

Correct: The computer is hers.
Correct: The typo was someone else's mistake.
Correct: Your computer should be personalized.

See the section titled "Common Word and Phrase Problems" for the use of *its* versus *it's*.

Apostrophes to show omission of numbers or letters. Although using the apostrophe to show omission of numbers or letters is too informal for most technical writing, we show it for completeness.

Wrong: Prestressing steel was introduced in the 20s.

Correct: Prestressing steel was introduced in the '20s.

Colons used to introduce a list, a clause, or quoted matter. The colon is most often used to introduce a list when the clause before the colon is an independent clause (i.e., a clause that can stand alone).

Wrong: Fixing the pump will require: an adjustable wrench, a socket set, and a few screwdrivers.

Correct: Fixing the pump will require a few items: an adjustable wrench, a socket set, and a few screwdrivers.

Colons can also be used to separate two independent clauses when the second clause amplifies or interprets the first.

Correct: The project manager told us the project was ahead of schedule: he said we had about 3 days' leeway.

Colons also are used to introduce a quotation related to an independent clause.

Correct: The sight of the Liberty Bell made her think of what Patrick Henry had said: "Give me liberty, or give me death."

Colons used in other ways. A colon is used in the salutation of a formal letter, in the representation of time, and to separate a title from a subtitle, as shown next.

Dear Mr. Petrosky:

We received the first draft of your book, *Practical Technical Communication: An Engineering Approach to Transmission of Ideas.* The meeting to discuss this draft will be on October 18 at 4:00 p.m. in the conference room on the fourth floor.

Commas, dashes, and parentheses to set off phrases. Additional information not essential to a sentence, often called *parenthetical information,* can be set off with commas, dashes, or parentheses, depending on how much you want to slow the reader. The previous sentence could also have been written as follows:

> Additional information not essential to a sentence (often called *parenthetical information*) can be set off with commas, dashes, or parentheses.

Most readers will read this sentence differently than the original sentence because the parentheses cause them to slow more than the comma does. Dashes have the same effect, but they also draw the reader's attention to the words in the phrase more than commas or parentheses do. Consider this sentence:

> Additional information not essential to a sentence—often called *parenthetical information*— can be set off with commas, dashes, or parentheses.

You, the writer, make the decision on the basis of how you want your reader to respond.

Commas in introductory phrases. A comma should be used after an introductory phrase unless the phrase is very short. What is "very short"? No hard and fast rule exists. You should use your ear. Read the sentence to yourself, pausing where the comma would be placed, then read it without pausing. One will likely sound better than the other. If not, use a comma. Look at the previous sentence. *If not* may seem like a relatively short phrase to you, but most people find that the sentence sounds better with a comma. Try it yourself. In this case, if you omitted the comma, doing so would probably not be considered incorrect.

Commas in lists. Using the *series comma,* the last comma in a list of items, will usually lead to more clarity in writing and is therefore preferred for technical documentation.

> Concrete is made from cement, water, sand, and aggregate.

Many people choose to omit the last comma.

> Concrete is made from cement, water, sand and aggregate.

The second sentence is correct; however, omitting the last comma presents problems if your list consists of items that are compounds. Consider this:

> John and Mary, Sam and Jane, and Rick and Sue are going to the diner.

In this case, each item is a couple, so *and* is used between pairs of names. Omitting the last comma produces this:

> John and Mary, Sam and Jane and Rick and Sue are going to the diner.

This sentence is not as clear as the first.

Commas vs. semicolons for linking two sentences. Two sentences connected with a conjunction, such as *and* or *but,* should have a comma *before* the conjunction.

> *Correct:* It is after 5:00 p.m., and we cannot reach town by dark.

Two sentences can be made into one without a conjunction by using a semicolon.

> *Correct:* It is after 5:00 p.m.; we cannot reach town by dark.

Two sentences can also just be two sentences.

> *Correct:* It is after 5:00 p.m. We cannot reach town by dark.

In general, what we referred to as *sentences* are technically *independent clauses.* Independent clauses should always be joined by a semicolon unless the conjunction *and* or *but* is used. This rule also applies when the second independent clause begins with an adverb, such as *besides, however, then, therefore,* or *thus.*

> *Wrong:* I have finished the problem, however, I am not sure it is correct.

> *Correct:* I have finished the problem; however, I am not sure it is correct.

Commas in single sentences with phrases and conjunctions. If the second phrase found after the word *and* cannot stand alone as a sentence, a comma should *not* be used before *and.*

> *Correct:* He is highly trained and is very competent.

When the second phrase is not a sentence and *but* is the connecting conjunction, a comma is acceptable but not necessary.

> *Correct:* He is highly trained, but is lazy.

Exclamation points. Unless you are quoting a statement with an exclamation point in it, exclamation points should not be used in technical writing.

Hyphens to join two or more words that form a single adjective. Hyphenation of words to form an adjective is debated in the writing community. Some writers use the hyphen in this manner all the time, and some use it only when it will prevent ambiguity. Use it when you believe it makes your sentence clearer.

> *Acceptable:* We need a six foot rod.
>
> *Better:* We need a six-foot rod.

> *Acceptable:* The time to failure plot is on the next page.
>
> *Better:* The time-to-failure plot is on the next page.

> *Ambiguous:* The engineer needs six foot long rods.

The preceding sentence could mean the engineer needs *six* foot-long rods or *six-foot-long* rods. In this sentence, a hyphen is required to prevent ambiguity.

Other uses of hyphens. A hyphen should be used to prevent awkward combinations of the same letter (e.g., re-elect, re-examine, and anti-intellectual). It should also be used to form compound numbers less than one hundred (e.g., *seventy-six* and *forty-four*). Last, it should be used to attach prefixes to proper nouns (e.g., *un-American, anti-Luddite,* and *ex-Chairman*).

Quotation marks with other punctuation. Quotation marks are used to set off direct quotations. Three rules should be followed:

1. Periods and commas always go inside the quotation marks.

   > *Correct:* The project manager said, "The project is ahead of schedule."

2. Colons and semicolons always go outside the quotation marks.

   > *Correct:* The project manager said, "The project is ahead of schedule"; he then discussed how to keep it that way.

3. A question mark or an exclamation point goes inside the quotation marks if it is part of the quotation and outside the quotation marks if it belongs with the rest of the sentence.

   > *Correct:* Who was it who said, "Give me liberty, or give me death"?
   >
   > *Correct:* The mechanic asked, "What happened to your air filter?"

Semicolons in lists. Semicolons should be used to separate list elements that contain a comma.

> *Wrong:* The company construction sites are in Chicago, Illinois, Fargo, North Dakota, and Madison, Wisconsin.

> *Correct:* The company construction sites are in Chicago, Illinois; Fargo, North Dakota; and Madison, Wisconsin.

Common Word and Phrase Problems

Accept and ***except.*** *Accept* is a verb with several meanings, including "to receive" or "to believe in". *Except* is also a verb, but it means "to leave out" or "to exclude". Although these words have vastly different meanings, they are often confused because they sound the same when we speak them.

> *Wrong:* I except that I will never truly understand calculus.
>
> *Wrong:* My classes are okay accept for Physics, which is extremely difficult.

> *Correct:* I accept that I will never truly understand calculus.
>
> *Correct:* My classes are okay except for Physics, which is extremely difficult.

Affect and ***effect.*** *Affect* is always a verb. In most writing, *effect* is a noun, but it will be a verb in certain cases.

> *Wrong:* Lead in gasoline effects the environment.
>
> *Wrong:* The affect of water and dirt on steel is well known.

> *Correct:* Lead in gasoline affects the environment.
>
> *Correct:* Lead in gasoline has a significant effect on the environment.
>
> *Correct:* The effect of the change was positive.

The following sentence shows an example of *effect* as a verb.

> *Correct:* To effect a change, we removed the director.

A lot and ***alot.*** No such word as *alot* exists; therefore, use of *alot* is always wrong.

> *Wrong:* Alot of things went wrong.

> *Correct:* A lot of things went wrong.

Among and ***between.*** In general, *between* is used for two items and *among* is used for many items.

> *Correct:* The choice was between stainless steel and aluminum.

Correct: A discussion among the design engineer, project engineer, and project manager resulted in the design change.

However, if you need to show the individuality of more than two items, you should use *between*.

Wrong: The choice was among stainless steel, aluminum, and titanium.

Correct: The choice was between stainless steel, aluminum, and titanium.

Amount and **number.** *Number* is used for countable items, and *amount* is used for bulk measure.

Wrong: The amount of screws used in this design seems a bit large.

Wrong: You are asking for a large number of money to do this work.

Correct: The number of screws used in this design seems to be too many.

Correct: The amount of concrete used in this design seems a bit large.

Correct: You are asking for a large amount of money to do this work.

You are probably more likely to use *amount* incorrectly since it often sounds appropriate even when it is wrong, but the incorrect use of *number* generally sounds odd.

As. The word *as* is overused and often used incorrectly by novice writers. The word has many proper uses; however, a few improper uses occur frequently. Among the worst culprits are *being as* and *seeing as*. These phrases should never be used in technical communication.

Wrong: The data values were entered in a file as opposed to entering each value individually.

Wrong: Operations were greatly reduced during the weekend, seeing as the pump was not working properly.

Correct: The data values were entered in a file instead of being entered individually.

Correct: Operations were greatly reduced during the weekend because the pump was not working properly.

The word *as* can mean *since* and *because*, but your documents will sound more professional if you refrain from using the word *as* in this way.

Okay: Model #5134 was selected for the data acquisition system as the other models would have cost more.

Better: Model #5134 was selected for the data acquisition system because the other models would have cost more.

Better and **best.** *Better* is comparative (two items), and *best* is superlative (more than two items).

Wrong: Of those two materials, steel is the best.

Wrong: After testing 10W, 30W, and 40W oil, we found 30W to be the better of the three.

Correct: Screws are better than nails in this application.

Correct: The best book of those three is the one by Jones.

Comprised of and **composed of.** *Comprised of* is always incorrect. *Comprise* means "include," so *comprised of* makes no sense. In general, use *composed of* in most sentences.

Wrong: The document is comprised of six parts.

Correct: The document comprises six parts.

Correct: Six parts compose the document.

Correct: The document is composed of six parts.

Concrete vs. cement. As an engineer, you should use *concrete* and *cement* correctly, even if most of the population uses them incorrectly. *Cement* is a material that holds other materials together. Thus, you have rubber *cement* to hold sheets of paper together. When you go to a lumber yard, you purchase a bag of *cement*. After you go home and mix the *cement* with sand, gravel, and water, you then have *concrete*. Asphalt and portland *cement* are used to hold sand and aggregate (gravel) together to make asphaltic *concrete* (often just called *asphalt*) and portland cement *concrete* (often just called *concrete*). So, a typical sidewalk is made from *concrete*, not *cement*.

Criteria is. Using *criteria is* is always wrong. *Criteria* is the plural of *criterion*, so you must use *criteria are*. Wait; we used *criteria is* in the last sentence. Isn't this construction wrong? No. When we used *criteria is*, the *is* referred to the *word criteria*, not to the actual criteria. The word *criteria* is singular. Oops; there it is again, but you get the idea.

Wrong: Appropriate design criteria is vital to the success of a project.

Correct: Appropriate design criteria are vital to the success of a project.

Correct: The tolerance criterion is difficult to meet.

Different from and **different than.** *Different from* is correct most of the time.

Wrong: My solution is different than yours.

Correct: My solution is different from yours.

One way to remember when to use *from* and when to use *than* is to write the sentence using the verb *differs* instead of the adjective *different*.

Correct: My solution differs from yours.

This sentence sounds correct and is correct. Replacing *from* with *than* will sound wrong and be wrong.

Wrong: My solution differs than yours.

In one particular situation, *different than* is acceptable: when it's followed by an adjective clause.

Acceptable: Cars are different than they were when I was young.

Few and less. *Few* (or *fewer*) is used for countable items, and *less* is used for items measured in other ways.

Wrong: I have less dollars than you do.

Correct: I have fewer dollars than you do; so I have less money.

Correct: The average worker should be able to complete this task in fewer than 5 hours.

Sometimes, *less* seems correct, and most people would not notice that it has been used incorrectly. This situation has once again resulted from changes that are gradually occurring in the spoken idiom of the English language.

Wrong: The average worker should be able to complete this task in less than 5 hours.

Although the preceding sentence is technically wrong, many people would find it acceptable.

Figured out. The phrase *figured out* is inappropriate in a technical document. Use a more formal word such as *determined* or *discovered*.

Wrong: I have figured out the solution to the problem.

Wrong: Machinists figured out how to make the manifold more efficiently.

Correct: I have determined the solution to the problem.

Correct: Machinists discovered a more efficient method for fabricating the manifold.

Infer and imply. *Infer* means "to draw a conclusion from information." *Imply* means "to suggest."

Wrong: I inferred in my memo that we should not bid on the job.

Correct: I infer from that noise that your engine has a problem.

Correct: I implied in my memo that we should not bid on the job.

Correct: Are you implying that I'm wrong? I inferred that from what you said.

Insure, ensure, and assure. *Insure*, *ensure*, and *assure* are often used interchangeably, and, in many dictionaries, they are listed as synonyms for certain situations. For technical communication, however, you should try to use them more carefully.

Correct: I assure you that I'll get the work finished on schedule.

In the preceding sentence, *insure* or *ensure* would be wrong. *Assure* means "to promise."

Correct: Seat belts will help ensure your safety.

In the preceding sentence, *ensure* is the best word, but *insure* or *assure* is acceptable to many people. *Ensure* means "to guarantee."

Correct: You should be insured against flood damage if you live on a flood plain.

In the preceding sentence, *assure* or *ensure* would be incorrect.

Irregardless. No such word as *irregardless* exists. Use *regardless* instead.

Wrong: The engine needs to be fixed, irregardless of cost.

Correct: The engine needs to be fixed, regardless of cost.

Is because. *Is because* is always incorrect. *Is* is a linking verb, and it needs a noun or an adjective on either side of it—for example, "The weather is hot" or "She is a professor." *Because* is the beginning of an adverbial clause, and you cannot use a linking verb with an adverb.

Wrong: The shift in data is because there was a sharp change in temperature.

Correct: The shift in data is due to a sharp change in temperature.

Its and it's. *Its* is the possessive of *it*. *It's* is a contraction meaning "it is."

Wrong: Its apparent that the power leakage has been reduced significantly.

Wrong: The graph shown in Figure 1 presents data from the experiment and it's slope is equivalent to those obtained in previous experiments.

Correct: It's critical that the test be conducted according to specifications.

Correct: The hypothesis test has been performed and its statistical significance determined.

Latin terms. Four commonly used Latin terms are found in technical documentation: *e.g., et al., etc.,* and *i.e.* These terms are defined in the following paragraphs.

e.g.   "for example"

The term *e.g.* is used when you want to give one or two examples of a given situation, but not necessarily an exhaustive list. The correct way to punctuate this term is to put a comma after the second period (similar to the way you would write *for example*, with a comma at the end of the phrase).

Correct: The engines in this class, e.g., model 5983, are subject to frequent breakdowns and require regular maintenance.

et al.   "and the others" or "and all the rest"

The term *et al.* is most commonly used when you are citing references with three or more authors. Also note that the correct punctuation for this term is a period after *al* but not after *et*.

Correct: Monte et al. stated that the design was achieved according to current ASTM standards.

etc.   an abbreviation of *et cetera* meaning "and so on"

The term *etc.* is used when you are making a list and want to show that other things could have been listed but you chose not to make the list too comprehensive. Note that the correct way to punctuate this term is to put a period at the end of it.

Correct: The choice of material for the gear was between a metal (steel, aluminum, titanium, etc.) and a plastic (polyethylene, polycarbonate, etc.).

i.e.   "that is"

The term *i.e.* is often erroneously used when *e.g.* is more appropriate. Use this term when you want to restate something, give the one and only example of a condition, or show an inclusive list. The correct way to punctuation this term is to put a comma after the second period.

Correct: The temperature was too warm to perform the tests; i.e., NIST standards could not be met.

Correct: The two most common structural steels were considered, i.e., A36 and A572 grade 50.

Quite and very. *Quite* and *very* should be avoided because they add nothing to a sentence except extra words. Simply rewrite your sentences without them.

Wrong: Titanium is quite strong.

Correct: Titanium is a strong material.

Very is more accepable than *quite,* but, like *quite,* adds little if anything to the meaning of a sentence.

Talked about. The phrase *talked about* is inappropriate in a technical document when you are referring to the *written* work of others. A better word is *discussed* or *presented.*

Wrong: The author talked about the use of computers in engineering.

Correct: The author discussed the use of computers in engineering.

Correct: The author presented data from several tests conducted at nuclear power plants.

The phrase would be acceptable if you were referring to a *speaker* and the document you were preparing was relatively informal. *Presented* would be a more formal choice.

Acceptable: The speaker talked about the use of computers in engineering.

Better: The speaker presented information about the use of computers in engineering.

That vs. which. *That* is used when a phrase is essential to the meaning of the sentence and should not be preceded by a comma; *which* is used when a phrase is parenthetical or nonessential and should be preceded by a comma. If the phrase you are considering using in your document needs a comma before it or commas around it, *which* is the appropriate word choice.

Wrong: The pump which is broken is in the main storage building.

Correct: The pump, which is broken, is in the main storage building.

Correct: The pump that is broken is in the main storage building.

In the first correct example, the pump is in the main storage building and, by the way, is broken. In the second correct example, the specific pump, the broken one, is in the main storage building.

The fact that. *The fact that* should be avoided because it adds nothing to a sentence except extra words. Simply rewrite your sentences without it.

Wrong: The fact that the engine is overheating indicates the cooling system has a problem.

Correct: The engine is overheating, which indicates the cooling system has a problem.

Their, they're, and there. *Their* means "of or relating to them," in which *them* can refer to people or things; *they're* is a contraction meaning "they are"; *there* is a place. Many other idiomatic uses of the word *there* have gained acceptance.

Wrong: Their going to work overtime to complete this project.

Wrong: They're discussion was productive.

Wrong: There work is often cited by other scholars.

Correct: Engineers must strive to improve their communication skills.

Correct: They're actively pursuing funding for their project.

Correct: Please, put the specimen over there.

Correct: We have enough time to go there and back.

Correct: Please put the report there on the table.

Too, to, and two. *Too* means "also" and "excessively," *to* has a wide range of uses and meanings, and *two* is a number. Generally, most people use the word *two* correctly but are often confused about the correct usage of *to* and *too*.

Correct: The results were verified by two separate and independent studies.

The use of *two* in the preceding sentence should be obvious, but incorrect usage of the word is common. The incorrect usage is probably the result of a missed typo since a spell-checker will not catch this error. A grammar-check program may catch it, but no substitute exists for careful proofreading.

Correct: The ambient temperature was too warm to conduct the experiments.

Correct: The second vessel ruptured along the welded seam too.

The two preceding sentences cover the primary uses of *too*. In the first sentence, *too* means "excessively," and in the second sentence it is used as a synonym for *also*.

The first sentence in the preceding examples also shows one of the common uses of *to*. In this sentence, *to conduct* is an infinitive. An *infinitive* is simply a verb preceeded by *to*.

Correct: The engine was pushed to the limit.

Correct: I will send the failed test specimens to the lab for further analysis.

Correct: Please come to my office to discuss the matter further.

Correct: Please give the data you collected to the other project engineers.

All the preceding sentences show correct uses of the word *to*. Note that the word *to* has many meanings and uses. (Look in a dictionary to get a better idea of these uses.) To use the word *to* correctly, you should probably think about using a process of elimination: Think about whether the word you want is either *too* ("excessive" or "also") or *two* (a number). If neither is the case, the word you should use is *to*.

Try and. *Try and* is always incorrect. You should use *try to*. (Another correct use of the word *to*.)

Wrong: I will try and solve that problem.

Correct: I will try to solve that problem.

Would of, should of, and could of. *Would of, should of,* and *could of* are always incorrect and are fairly common mistakes made by novice writers. Such mistakes are usually made because *would of, could of,* and *should of* are frequently part of the spoken idiom, in which people, through time, have contracted the word *have* to the word *of* (they sound somewhat similar if you say them rapidly). The *of* in each of these three phrases should be *have*.

Wrong: I could of completed that analysis for you.

Correct: I could have completed that analysis for you.

Your and you're. Another pair of synonyms that seem to cause a great deal of problems for novice writers is *your* and *you're*. *Your* means "of or relating to *you*," and *you're* is a contraction meaning "you are."

Wrong: Your not conducting the experiment according to ASTM standards.

Wrong: You're colleagues have commended you for a job well done.

Correct: Your study habits will play a significant role in your ability to perform well at the university.

Correct: You're going to be working on this project for the next few months.

When trying to decide which of these two words to use in a document, ask yourself if you could replace the word by *you are* without changing your meaning. If so, use the contraction *you're*; if not, use the word *your*.

Other Considerations

Lists: parallel structure. Lists can be used within a paragraph or can be separated from the paragraph as bulleted or numbered lists. A list consisting of single words or only two or three items may be placed within a sentence in a paragraph. This type of list is referred to as an *embedded list*. A list that is more complicated than a list of single words or a few items should be separated from the paragraph. Also, if you think the items in a list should be numbered or you want to emphasize the list, then, generally, separate it from the paragraph. The two types of lists that are separated from the paragraph are a *bulleted list* and a *numbered list*. Examples of lists follow.

■ *Embedded lists.* Lists of a few items or lists of single words may be placed within a paragraph. Use a colon before the list unless the last word before the list is *are*. *Examples:*

The materials that could be used for the fitting are steel, aluminum, or titanium.

The following materials can be used for the fitting: steel, aluminum, or titanium.

The options are renovating the building, tearing it down and rebuilding it, or leaving it as is.

We have the following options: renovating the building, tearing it down and rebuilding it, or leaving it as is.

Note that no items in the list were capitalized. In embedded lists, capitalize only the items that would ordinarily require capitalization, such as proper names. *Example:*

The following companies are qualified to perform the work: Acme Engineering, Select Engineers, and Superior Engineering.

■ *Lists with bullets.* A bulleted list is appropriate if the items in the list do not need to be in any specific order. The lists in this section are all bulleted because no single item is more important than another and no single item must follow another. The list under the *numbers* section and this list are good examples of bulleted lists.

■ *Lists with numbers.* Use of a numbered list is appropriate when you want the items in the list ordered, either for preference or for organization. For example, a list defining a procedure should be numbered because the first step should be done first, the second step second, and so forth. A list of preferences should be numbered if the items are in order of preference from most desirable to least desirable. Another option would be to use a bulleted list with a statement that the list is in preferential order, but a numbered list with the same statement is more clear. *Examples:*

Our team considered three options for your building. The options are listed below from most highly recommended to least recommended:

1. Renovate the building.
2. Demolish it and rebuild.
3. Leave it as is.

The list in the preceding example could be created in any of the three ways. The numbered method makes the preference clear and emphasizes the list. Each item has a period at the end because it is a sentence. The general rule is to place a period after each item if *any* one of them is a sentence, but do not place a period after any of them if all of them are phrases. Thus, the preceding example could be as follows:

Our team considered three options for your building. The options are listed below from most highly recommended to least recommended:

1. Renovation
2. Demolition
3. Status quo

In bulleted and numbered lists, all items should begin with a capital letter, as shown in the examples.

Lists, whether embedded or not, should have *parallel structure,* which means each item in the list must have a similar grammatical structure. Consider the preceding list:

Renovate the building.

Demolish it and rebuild.

Leave it as is.

All items in the list begin with a verb (i.e., the list has parallel structure). However, consider if the list had been written as follows:

Renovation of the building.

Demolish it and rebuild.

Leave it as is.

In this case, the first item begins with a noun and the other two begin with a verb. This list does not have parallel structure.

Numbers. Generally, numbers less than 10 are written in words, and numbers greater than or equal to 10 are written as numerals. Some exceptions to these general rules follow:

- Use numerals for numbers with units attached. *Examples:* 5 m, 6 lb, $12 per hour.
- Use words to write a number that begins a sentence. *Example:* Fifteen workers were injured in the collapse.
- Use numerals for numbers in a group when some numbers are less than 10 and others are greater than or equal to 10. *Example:* Of the 15 injured workers, 10 are in good condition and 5 are in critical condition.
- Use numerals and words when two numbers are used in succession. *Example:* The order was for twenty 16-foot pieces of lumber.
- Use numerals with dollar amounts. *Example*: The pump cost $5,000.
- Use words for ordinals. *Example:* The first, second, and third people can now be let through the gate.
- Use commas for numbers with four or more digits. *Examples:* 5,000; 1,034,444.

Pronouns. Following is a list of rules for pronoun usage.

- A pronoun must agree with its antecedent.

    Wrong: The company chose their design team last week.

    Correct: The company chose its design team last week.

    Correct: The members of the union elected their steward.

- The antecedent of a pronoun must be clear.

    Unclear: The city engineer's assistant said she is not available this afternoon.

    Clear: The city engineer said she is not available this afternoon.

- The word *this* by itself should generally not be the subject of a sentence.

    Wrong. This is an example of using *this* incorrectly as the subject.

    Correct. This sentence is an example of using *this* correctly as part of the subject.

In the first sentence, the word *This* at the beginning of the sentence is acting as the subject of the sentence and what it is referring to is possibly unclear. The second sentence, in which *This sentence* is the subject, is much more clear.

Repetition. Do not use the same word multiple times in a sentence unless doing so cannot be avoided. Rewrite the sentence to eliminate multiple use of the same word.

Rhetorical questions. In general, use rhetorical questions infrequently, and in technical reports avoid them altogether. A *rhetorical question* is asked to lead into a discussion, with no expectation of an answer from the reader or listener. (*Example:* Why did we choose titanium over steel in the design?) In formal technical documents, rhetorical questions should be avoided, but they are acceptable in some informal documents. They are also acceptable in presentations. In presentations, they can provide a good transition to another idea, but don't overuse them because they will lose their effectiveness.

Spelling and grammar. Most word-processing software provides tools for checking spelling and grammar. Take advantage of them. At the same time, you should proofread the report to check for errors the spelling and grammar tools may have missed. For example, careful proofreading is necessary to find errors such as using *to* instead of *too*, or vice versa; *there* instead of *their*, or vice versa; or *its* instead of *it's*, or vice versa, as discussed previously.

Split infinitives. In English, unlike most other languages, the infinitive form of verbs includes the word *to*. Examples of infinitives are *to eat, to be, to walk, to run*, and so forth. When writing, you should always try to keep the two words in the infinitive together. Inserting a word between *to* and its verb is called a *split infinitive* and should be avoided when you are writing. The difficulty is that many times the split infinitive will sound better or prevent awkward sentence construction. In fact, if you sometimes split an infinitive, most people will find it acceptable; however, you should try to avoid it when possible.

Wrong: The processor was not able to efficiently handle the data.

Wrong: A bar chart was chosen to more clearly show the trends in the data.

Correct: The processor was not able to handle the data efficiently.

Correct: A bar chart was chosen to show the trends in the data more clearly.

Note that the two "wrong" sentences are wrong only if you are meticulous about split infinitives. Some writers would argue that *to efficiently handle* in the first wrong sentence is good structure because *efficiently* clearly relates to *handle*. A similar argument could be used for the second wrong example. We believe you should not split infinitives unless you have a good reason to do so and can defend your reason. Certainly, some cases of split infinitives sound good: "To boldly go where no man has gone before."

Verb tense. You should use verb tenses in your document that are consistent with the occurrence of each event in a paragraph. The types of verb tenses are described next.

- *Present.* Action occurring now.

    Example: The engineers design the car.

- *Past.* Action occurring in the past.

    Example: Last year, the engineers designed the car.

- *Future.* Action occurring in the future.

    Example: Next year, the engineers will design another car.

- *Present perfect.* Action continuing up to now.

    Example: During the last 9 months, the engineers have designed the car.

- *Past perfect.* Action continuing up to a specific point in the past.

    Example: Before the end of last year, the engineers had designed two other cars.

- *Future perfect.* Action continuing to a specific point in the future.

    Example: By the end of next year, the engineers will have designed four cars.

Consider the following paragraph:

Acme Construction Company *loses* a significant number of contracts. This problem *appears* to be a management issue. During the past 2 years, it *has lost* 140 of the 150 jobs that it bid on. Last year alone, the company *lost* 90 jobs. Prior to 2 years ago, the company *had won* 20 percent of its bids. Now Acme *wins* about 7% of its bids. With the proposed change in management, the company *will increase* its bid success rate. By the end of the second year, the success rate *will have increased* to at least 20%.

This example is extreme since we placed all six verb tenses in one paragraph, but you should be able to see how each tense fits. The decision about which tense to use must be made carefully so that you make a sentence or paragraph mean what you want it to. Consider the second sentence:

This problem *appears* to be a management issue.

Change the verb tense so that the sentence reads as follows:

This problem *appeared* to be a management issue.

The meaning of each of these two sentences is noticeably different: In the first case, when present tense is used, the problem currently exists. In the second case, when past tense is used, the problem existed in the past and may no longer exist.

Wrong verb tense usage is common in student design or experimental reports. (These types of reports are covered in Chapter 6.) In design and experimental reports, you will almost always write about something you already did. The action occurred prior to the writing, so any sentence that explains the experimental procedure you used or the design choices you made should be written in the past tense. However, generally your results are still true and should be stated in the present tense.

Wrong:	Fifty grams of the compound is weighed and placed in a beaker.
Wrong:	This design alternative was selected because it seems the most cost effective.
Wrong:	Poisson's ratio of the material was 0.33, which differed by only 5% from the reference value.
Correct:	Fifty grams of the compound was weighed and placed in a beaker.
Correct:	This design alternative was selected because it seemed the most cost effective.
Correct:	Poisson's ratio of the material is 0.33, which differs by only 5% from the reference value.
Correct:	Poisson's ratio of the material was determined to be 0.33, which differs by only 5% from the reference value.

The use of the correct verb tense is vital to the meaning of your document. You must take the time to carefully consider the time frame for the occurrence of the action you are describing and compare it with the current time when you are writing the document.

Index